Moral Dimension of Man in the Age of Computers

Cognitive Disability
& its challenge to Moral
Philosophy
Ed. Eva Feder Kittay
& Lilian Carlson

Adam Drozdek

University Press of America, Inc.
Lanham • New York • London

Copyright © 1995 by
University Press of America,® Inc.
4720 Boston Way
Lanham, Maryland 20706

3 Henrietta Street
London, WC2E 8LU England

Library of Congress Cataloging-in-Publication Data

Drozdek, Adam
Moral dimension of man in the age of computers / Adam Drozdek.
p.   cm.
Includes bibliographical references.
1. Man. 2. Ethics. 3. Philosophy and science. I. Title.
BD450.D78   1995   170 --dc20   95-16818 CIP

ISBN 0-8191-9983-4 (cloth: alk: paper)
ISBN 0-8191-9984-2 (pbk: alk: paper)

⊖™The paper used in this publication meets the minimum
requirements of American National Standard for Information
Sciences—Permanence of Paper for Printed Library Materials,

Contents

Introduction

The essence of teaching of logical positivists - says Thomas Merton - can be summarized in the following statement: "since we cannot really say anything about anything, let us be content to talk about the way in which we say nothing." It is a result of "the mechanical clicking of the thought machine manufacturing nothing about nothing, as if even nothing had at all costs to be organized, and presented as if it were something." Logical positivists "don't have the imagination or the good sense to stand in awe at real emptiness. In fact, their rationalizations seem to be a complacent evasion: as if logical formulas somehow could give them something to stand on in the abyss. And now: just wait until they start philosophizing with computers!" (1966, 8). And also about computers! These rationalizations of today's positivists seem to be substantiated by the incredible strides of computer technology and virtually unbound applications of computers. Philosophizing seems to unfold before our own eyes as the fact of technological progress defying any barriers. The computer becomes the tool and model, the vehicle of progress, and the foundation of philosophical reasoning. It acquires life of its own to the extent that it is said to possess characteristics not ascribed before to inanimate objects, not even to very life. They are said to break out from the realm of ordinary tools to the domain reserved up until now only to men. Computers are being vivified and endowed with personality. They are not mere tools any more; they are on its way of becoming part of society with their rights and obligations. Computers emerge as partners and companions, and should be treated accordingly.

This high status of computers is due to the reasoning abilities they possess; not to emotions or affections, but to their cognitive characteristics, and also to their speed and precision. This rationality, logical rationality, unswervingly followed and very rigid (even if only heuristics are applied), is the core of computers, and there is little need with endowing them with other abilities to ensure their efficiency. Rationality is the key to their success, this rationality, which for centuries has been held in the highest esteem by many. The priority and predominance of the rational dimension, therefore, is proven practically by computers and by their role in today's society. Praxis is the proof that the rational dimension ought to be given priority over the other dimensions characterizing man: religious, moral, aesthetic, affective, etc. Praxis is also the proof that if this rational dimension is extracted from man and put in pure form in computers, then progress in general is guaranteed, in particular, technological progress. Hence, other dimensions of man should not be given too much weight, otherwise, the path of progress may be unnecessarily hampered. This is the lesson which we seem to be taught.

Computers were modeled on the rational dimension of man, on cognitive abilities manifested especially in proving theorems or in playing chess. This leads to seeing personhood in computers, that is, to showing that rationality appears to be constitutive to personhood to the exclusion of other dimensions. This, in turn, leads to elevating the rational dimension of man to the highest status with little attention given to other dimensions. "Man is primarily a rational being" shifts toward "man is only a rational being." If not explicitly stated, such an attitude can be seen in today's cognitive psychology, which is enchanted by computational model of man in which there is only insignificant room for other dimensions. And this is what this book is against.

The main claim of this book is that defining man in terms of the rational dimension is invalid. Man is primarily a moral being, not a rational being. It does not mean that all men behave morally, at least not at all times. It means that the moral dimension is constitutive to man and it is primarily through this dimension that man is considered human. The rational dimension, albeit important and indispensable, plays only a subsidiary role to the moral dimension. It is an efficient and sophisticated tool used by the moral dimension to achieve its goals, hence ascribing a predominant standing to the rational dimension in human life is at the very least insufficient.

The rational dimension is concerned in truth, the moral dimension concerned with good. The moral dimension can have (at least) two, oftentimes interrelated forms, depending on the way good is understood. Good can have a metaphysical standing and be understood as a part of the universe; it can be a supreme good, God, or the idea of good. The primacy of the moral dimension would mean striving for knowledge of this good or for being united with it. Good can also be understood as goodness, other-centeredness, and altruism. With such an understanding, moral dimension is more earth-bound and oriented toward other people. It is man's nature to live for others and with others - even if this characteristic is suppressed during his lifetime. Living for others may be only a stepping stone for attaining the supreme good, a way for being united with the Good.

Regardless of the way good is understood, man needs to act purposefully and efficiently to attain it, and hence, man needs a cognitive dimension. Man is characterized by the moral dimension first, and only secondarily by the rational dimension. The argument of this book is that this priority became unfortunately reversed due, among other things, to the fascination with computers. Guilty of this fascination have been primarily scientists themselves, which was and is manifested in some of their farfetched claims. Time already has proven many of these claims to be at least exaggerated.

In the first chapter it is shown how developments in artificial intelligence - the most philosophically oriented branch of computer science - led to giving the rational dimension its overrated status. As a consequence of this status, computers are viewed as a form of life, as rational beings endowed with personality, which raises the problem of computers having rights. The first chapter takes issue with this idea. The second chapter presents a discussion of the animal rights movement which also wants to include animals among men as equals with proper rights. The third chapter shows that the claim of this book - man is primarily a moral being - is by no means new and it can be found in the views of philosophers. The fourth chapter draws on moral development research conducted by many psychologists to show its invariants. This view is developed in particular by cognitively oriented psychologists, who stress very strongly the role of the cognitive dimension in morality. This view is insufficient and it should be amplified to include the affective side of man in the picture. Many psychologists today rediscover this dimension of man and its role in moral development. The last chapter is a discussion of a role of school, in particular, the university in reestablishing the position due to moral dimension.

References

Merton Thomas, *Conjectures of a guilty bystander*, Garden City: Doubleday 1966.

Acknowledgments

I would like to thank the editors of *AI & Society* for their kind permission to include an amended version of "Moral dimension of man and artificial intelligence", copyright © 1992, Springer Verlag London Ltd., as Chapter I, the editors of *Journal of Information Ethics* to allow me to include a revised version of "Pecunia non olet?", copyright © 1992, McFarland and Co., Inc., Publishers, Jefferson, NC 28640, in Chapter V, and the editors of *Social Epistemology* for their permission to include fragments of "The possibility of computers becoming persons", copyright © 1994, Taylor & Francis Ltd., in Chapters I-V.

Chapter 1

Moral Dimension of Man
and
Artificial Intelligence

Science is one of the greatest achievements of the human mind. It attempts to capture, through a system of concepts and rules, the regularities that underlie a course of all phenomena in the universe. Science is a realm of reason and rational reasoning that constantly refers to perceptual data. It uses the most rigid formal apparatus possible along with a stringent logic to derive statements that can be confirmed or disproved by experiment or observation. But there has always been a vast domain which science was unable to capture by its theoretical apparatus. This domain, however, is also fascinating to the human mind, and an insight into it is absolutely vital for an adequate functioning of individuals in society. Therefore, philosophy can always flourish, ethics has a good reason for existence, and aesthetics is needed. These domains are less precise in their statements and less structured than science, whereby science can claim priority with regard to its achievements. Most of the time science made an effort to impose an image it generated as a valid representation of the universe. Thus, from its inception, science was associated with reductionism, such as Pythagoreanism, mechanicism during the era of the Industrial Revolution, physicalism of the beginning of the 20th century, and today's sociobiology or programmabilism (Drozdek 1990; 1994). Science has always created a more or less orderly picture of a certain domain and tried to enlarge it to refer to the world in its entirety. Science all along attempted to enforce looking at the world from a particular perspective as the only, or at least primary, way of seeing the world, and gained an upper hand over other domains, such as religion, common sense, or art, since it was the best organized and most cohesive. In all the reductionist perspectives mentioned above, ethical issues occupied very little room, and if possible, they were reduced to nil. For instance, in physicalism they were treated as pseudo-problems, and in program-mabilism they are reduced to the status of decisions expressed by if-statements.

Science has always emphasized the importance of rational cognition, and using reason as a last arbiter in epistemological issues. Perception, to be sure, has always been included in the picture, but perceptual data without the organizing power of reason were unable to create science. Pure sensualism as an epistemological position in scientific methodology is simply untenable. Thus, science stresses the role of reason and logical reasoning as the means of creating theories, offering explanations, and propounding hypotheses. The problem of good and evil is extraneous to science, and ethical issues do not belong to science proper. Therefore, there is only one step to the promul-gation that these issues, along with pondering the problem of good and evil, are an outright hindrance to the progress of humanity and a hurdle standing in the way of man's development as a truly rational being. This position is exemplified by statements from Dean E. Wooldridge.

Wooldridge envisions the future world as a place in which the mechanical image of man will gain a general acceptance. Is such a "society of machines" in any way endangered by attitudes found in the present world? Can this society use in any way religion, law, and morality? Yes, in a limited way, and as a necessary evil. Religion will be retained only for those who may wonder about "the seemingly inexplicable origin of the laws and particles of physics". Punishing a transgressor of the law so that the punishment contributes "most to the good of society" is, according to the mechanist, pointless. And most of all, "the disappearance of the mystical concept of right and wrong could significantly increase the logical content of human thought" (Wooldridge 1969). For the concept of right and wrong unnerves the power of logical reasoning, obstructs the ability of seeing a subject matter clearly, and unnecessarily intervenes in the endeavor of expanding the realm of cognition. Hence, to be more human and more humane means to be more rational and less imbued with the obsolete problem of right and wrong.

"Society profits - continues Wooldridge - when its members behave more intelligently. And men who know they are machines should be able to bring higher objectivity to bear on their problems than machines that think they are Men." First, it has to be observed that the concept of "higher objectivity" is as mystical as the "concept of right and wrong". It simply presumes that it is possible to find a yardstick of objectivity stating how close our knowledge is to the subject matter. It is an old delusion, so clearly expressed by the concept of the protocolar statements of the positivists. There is no pure, objective cognition, therefore no independent yardstick can be found that measures our nearness to the *noumena*, the things in themselves. Secondly, what does it mean to behave more intelligently? More efficiently, faster, more productively? How is this intelligence measured? What are criteria of intelligence? It is doubtful whether society profits if some perpetrators "behave more intelligently" than their predecessors. Besides, when it comes to a decision of where to apply certain scientific results, what would be the most intelligent decision? It depends, to be sure, to whose benefit the decision is made, that is, who makes the decision. Unless these people are literally mechanical and conditioned somehow after the fashion of the brave new world, no unanimity can be expected, and thus some conflicts of interests will arise. Hence, interests of a given group would delimit a more intelligent behavior.

Wooldridge's pronouncements are probably extreme, but by no means isolated. The astounding technological progress of our century creates an atmosphere of giving an ultimate priority to reason, thinking, and faculties of mind. Science is being expanded thanks to the rationality of man. Science progresses technology, and thereby our well-being, *ergo* everything that is, or rather, everything that counts, that is progressive, modern and important, is gained through the power of logical thinking and relying upon the power of reason. Therefore, if contemporary futurists see some crossroads, some open ends and different possibilities for the future of the society, they resort to the power of reason as the only human capability that ensures making the right choices. "The wealth we seek - writes Jean-Jacques Servan-Schreiber - does not lie in the earth or in numbers of men or in machines, but in the human spirit. And particularly in the ability of men to think and to create." Also, "the training, development, and exploitation of human intelligence - these are the real resources, and there are no others" (1979, 240-1). It implies that trying to exploit other human faculties is out of place, since they can hardly be considered real resources. Hence, if they cannot be included in the category of such resources, they are either harmless vestigial organs of the pre-scientific past or obstacles in the way of progress and, as such, should be surmounted or simply removed. It also implies that moral considerations have little, if anything, to do with progress, and therefore humanity should not refer to them when trying to envision the future paths of society. The future can be conjured up by the power of reason, not by a recourse to moral values. It is even inappropriate to use the latter in the context of depicting the future. Reason, intelligence, thinking - that is all that counts.

Therefore, man is split into two parts, with his true and progressive part located in the brain, and then everything else. Today, the brain is mainly interested in efficiency and, occasionally, as a means, in the classical virtue of truth. The extra-rational part revolves around the virtues of good and beauty. The latter part is allowed to exist if restricted to individual affairs and private life; the former reigns at a larger scale as the only virtue sufficiently keen and perspicacious to tackle all the serious problems of society. The rational part of man is given an upper hand; the religious, moral, aesthetic, etc. are either merely tolerated or outright suppressed. In the name of acquiring true humanity, man is stripped of what goes beyond the rational and restricted to his brain. Technological progress rooted in the rational feeds it and wants the non-rational to wither away. Today, this progress would not be possible without the computer; therefore, the image of man painted by computer science, especially by artificial intelligence (AI), is of great significance, because, except for biology, no other domain contributed so much to this one-dimensional image of man.

One of the first areas of artificial intelligence was theorem proving. The researchers were captivated by an axiomatic method used in logic and mathematics that allowed one to derive an infinite number of theorems from a set of axioms and a set of inference rules. Russell and Whitehead's *Principia mathematica* (1910-13) was a model of this approach: Newell and Simon's *Logic Theorist* (1957), and then the *General Problem Solver* (1959) were systems designed to show that the computer was able to prove some theorems of propositional logic and first-order predicate calculus. The work on these systems led to the development and formalization of several searching and planning strategies (to mention only the means-ends analysis). A further interest in theorem proving materialized in devising the resolution technique that, in turn, led to the creation of logic programming language, Prolog (or rather Prolog interpreters).

Another old area of artificial intelligence is game playing. The earliest, and still very interesting, is Samuel's checkers player (1959), which could improve its performance by changing weighing parameters in its evaluation function according to the situation on the board. Then there are many chess players. All these systems are continually being improved. In fact, computer checker players are able to beat human champions, and although chess players cannot do that, their performance is truly amazing.

What is interesting about these first domains of AI is their exclusiveness; very few people are interested in proving theorems, and even less do that professionally. Similarly, the game of chess is not a common game in the sense that everyone plays it. Although basic rules can be learned in a matter of a day, decent, let alone very good, playing is acquired only over a matter of years. There is, therefore, an aura of specialness about these areas, and a conviction that higher intelligence and unusual skills are needed to become initiated in them. Although it is true, this way of picturing the skills of theorem proving and playing chess usually associates intelligence with this sort of ability only at the cost of refusing to ascribe any intelligence to more mundane skills such as preparing a dinner or walking down the street. It says that an intelligent person can play chess well and do mathematics, and whosoever does it only to a limited extent or is unable to do that is unintelligent.

An original rationale for the AI research was to acquire better knowledge about man and his abilities and to uncover mental processes leading to certain results. For instance, the work on *Logic Theorist* was motivated by an attempt "to understand how a mathematician, for example, is able to prove a theorem even though he does not know when he starts how, or if, he is going to succeed" (Newell, Shaw, Simon 1957, 109). Also, *GPS* was assessed as a step toward a theory of heuristic which can be used "both to understand human heuristic processes and to simulate such processes with digital computers" (Newell, Simon 1958, 6). "An attempt to build working models of human behavior ... requiring the machine's behavior to match that of human subjects" generally characterized the first AI projects (Minsky 1969, 7). The goal was to build a machine that can repeat human feats by perceiving, thinking, and acting the way humans do. To that end, researchers observed how human subjects solved problems and took protocols of their thinking aloud to employ them in designing *GPS*.

Although it has already been presumed that man was just an information processing system, a motivating force of the research was to know more about human nature, even if only within the confines of the information processing model.

In these attempts, the researchers were interested in what seemed to be unusual achievements of the human mind, that is, in proving theorems or playing chess, aiming at what seemed to be human *par excellence*. Initial successes were very encouraging and, in a sense, intoxicating: the researchers attempted to apply the same computational treatment to other human abilities that are not so unusual as those, whose mystery seems to have been unraveled by devising a particular set of heuristic techniques. If achievements of the greatest chess players and logicians can be approximated by an information processing model, it should be even easier to mimic computationally other not so impressive skills that all humans share. Yet, the situation substantially worsened when researchers became interested in these skills that seemingly did not require much intelligence, such as speaking, hearing, seeing, walking, learning, understanding speech, etc. Even babies are able to recognize faces of their mothers, children at a very early age are able to speak and understand, and children reared in a bi-lingual environment can with little difficulty switch languages and translate. Not surprisingly, these skills did not catch the attention of the first AI researchers. However, after they became a subject of research, all attempts to formalize them and transfer them to the computer met with at most partial success. For instance, there are many systems for machine translation, and many numerous groups are working on them in different countries. Yet Yehoshua Bar-Hillel's criticism concerning some early systems of this type still applies to what we can see today, and no system can pretend to live up to the title of the fully automatic high quality machine translation system.

Natural language processing developed into a very broad field, but its rapid progression slowed down somewhat after it was understood that relying on syntax alone would not suffice. The reason for using any language is to convey some information, to communicate something, that is, what is the most important about language is its semantics. However, semantics can be understood only in connection with the knowledge of the speaking person, especially common sense knowledge. And here is where insurmountable problems arose, namely in an attempt to represent knowledge. The researchers developed some methods for knowledge representation (semantic networks and frames) and created working systems, but these systems were restricted to very specific domains and by no means measured up to human performance.

With all these problems besetting the researchers, the underlying rationale behind the AI research changed. Initially man and human abilities were looked up to in order to construct a machine. The computer was to work according to what is known about man and his high level skills. Since workable systems were created, the whole of man was pushed into the Procrustean bed of the information processing model. Not only cognitive processes were treated this way, but also other human faculties, such as feelings, emotions, intuition, creativity, intentionality, consciousness, beliefs, sense of humor, conscience, etc. If these faculties did not suit themselves to the computational approach, it was an argument against them rather than against the approach itself. From here there is just one step to a declaration that they are unimportant, and thus can be ignored or simply said to be redundant categories of old ages. For instance, we can read an assertion that "the presupposition that consciousness is an important, or a real, phenomenon should be dropped" (Wilkes 1984, 241) and that we finally ought to cease to "protect the ghostlike existence of the free will, [since] contention of its existence really makes little difference to the AI projects" (LaChat 1986, 289). These characteristics were not treated as an integral part of man, but as elements disturbing the proper functioning of man as a cognitive agent. They could not be transferred to the computer, *ergo* they were irrelevant. Even if they could be incorporated into the computer system, it would be done out of fancy rather than a real need. In this way, the computer became a model of deciding what is to be human in man. Instead of modeling computer after man, the computer itself became a model of humanness.

When constructing this computational picture of man, very little attention was paid to such problems as moral or aesthetic values. Although moral choices could be simplistically understood as a set of decision statements and as such incorporated into a computational model, no attention was paid to them. Man was presented as an intelligent being; so intelligent, in fact, that what could not be considered intelligence was neglected or even denied existence.

The rational dimension of man was singled out as a constitutive characteristic of the species as reflected in its name, *homo sapiens*. However, man is not only a thinking being, a being who uses its brain to survive, but also a being characterized by other dimensions: the moral, the religious, and the aesthetic. Only man is able to tell between good and evil, only man is able to admire beauty. The humanity of *homo sapiens* is summarized in the Platonic triad of truth, good, and beauty. Only man has the capacity to truly perceive these virtues, and if these dimensions are severed from one another, *homo sapiens* ceases to be *homo* and becomes *res sapiens*. Using rational capacities certainly gives man an advantage over other known beings on the path of survival, but using only this capacity by curtailing the moral and aesthetic dimensions reduces him to the level of thinking, animate machine.

The aesthetic and the ethical aspects are as constitutive of man as the rational. Retaining the latter when neglecting the former is useful for some purposes, for instance, to have a better insight of human thinking processes, but it is unacceptable as an ultimate model of man as a member of society. But it was exactly the tendency of AI, and now it is a model offered by cognitive science that bases its research on the model of man once created by AI. It is very unfortunate that this model spilled over to contemporary psychology and has been embraced so readily by it. Using only such a one-dimensional model of man, psychology somewhat overlooks the *psyche* part of its name and overemphasizes the *logos*.

Another reason for this disinterest in what goes beyond cognitive processes, including decisions of moral nature, has been indicated by Steve Torrance. He says that the work on "producing computational models for emotions, aesthetic and evaluative attitudes, choices, etc." is neglected "because of its relatively low payoff in military and commercial terms." The usefulness of any "computational ethical modelling" is doubtful, since "the main volume of research in the field of AI is becoming more and more subservient to the needs of developing advanced systems of weaponry and warfare, to the accelerated accumulation of wealth and power by multinational industrial concerns, to the battles for world market domination" (Torrance 1986, 70).

In 1983 DARPA launched the Strategic Computer Program, a research project designed for developing intelligent machines for military application. The project lists three specific applications: autonomous vehicles, expert associates for pilots, and a large-scale battle management system. The Project is to develop new hardware and software with a focus on AI, especially on speech recognition, computer vision, natural language processing, and expert systems. The Project is expected to generate some applications to be used by industry. The primary goal, however, is to produce more weaponry, more sophisticated armament and deadlier destruction tools. Industrial spinoffs, in fact, are to strengthen the efficiency of military applications, thereby expanding the weapon producing industry. Millions of dollars are allocated to the SCP budget each year to produce more intelligent and powerful weapon systems. The universities and research centers are drawn into SCP to conduct AI research, becoming subservient to the needs of the military.

In this situation, it would be surprising to see the AI researchers interested in ethical issues and in building a model with a possible application in machines. Would it make much sense to implant any moral values into highly intelligent machines designed for killing? Such values would only become an obstacle in the high performance of such machines. Ethics and battlefield can hardly be mentioned in the same sentence, but much less become a part of the same contrivance.

Besides, AI is no longer interested in philosophical issues with regard to the nature of the human mind; it has become an applied science trying to produce workable systems for military and industrial applications. The original rationale of building a computational man is not appealing any more. The focus is on machines and systems that can be efficiently applied in some domains. Certain functions of these systems resemble intelligent human behavior, but this is irrelevant, and the emphasis is shifted to the efficiency of the product. If systems based on the model of a sea gull turned out to be more efficient than those trying to mimic human abilities, AI research would make the sea gull its favorite research object, and any attempt to know man better by using the AI methods would be abandoned. When AI lost its philosophical flavor, it became a tool factory driven by fund sources. Since the sponsors are the least interested in ethical issues, so is AI. As a result, we are left with a crippled, half-made model of computational man created by AI decades ago. And yet because of the prestige of computer science, the model is most widely recognized as an official model of our epoch.

Unfortunately, AI did not pay much heed to the warning spelled out by Margaret Boden to the effect that "psychological and social theories in general are not purely *descriptive*, but largely *constitutive* of social reality, and computational theories are no exception. The role assigned to human beings in the popular social theory is liable to become the role ascribed by individuals to themselves" (Boden 1977, 471). Since man is merely a rational being, so is the human society. Thus, what cannot be placed within the confines of rationality is undesirable. In this way a popular social theory can be created and validated by the scientific authority of the field that has borne it. This authority is sufficient for society to accept the theory and make it reality.

According to this social theory, the recent resurgence in AI of connectionism, once pronounced to be "without scientific value," is not revolutionary. This resurgence was caused by the inmarketability of systems based on the computational model and not by a philosophical interest in modeling man, as it was before suppressing it two decades ago. The model may change, but the emphasis remains the same: only the rational part of man is taken into account, while the other sides are barely noticed or simply avoided. Whichever model is taken into consideration, it very closely resembles the ideal man of the future Wooldridge depicted: a purely rational being with the moral side expunged, consciousness expurgated, sensations wiped out, feelings eradicated. There remains a purely rational man who finds his ultimate fulfillment in sophisticating armaments.

The undeniable efficiency of computer systems is also an argument in favor of the model of man: if this model served as a platform for creating these systems, and the systems are so efficient (in most cases), then the model must be true. The weapons produced are hitting the target much more precisely than before, they can intercept missiles much more efficiently, and they are smaller, lighter, and yet more powerful. What other proof does one need to confirm an underlying model? If some ethical aspects were included in this picture, hence in these weapons, they would become more hesitant, more reflective, and less efficient, thus less desirable. Therefore, the same should be done with man, presumably only to his advantage. In this way, a simplistic model of man is presented as sophisticated, and a far-fetched reductionism as the true way. The image of man as *only* a rational being acquires scientific validity and patronage of the market. The voice of ethicists can be neglected as simply irrelevant. If one has to choose between philosophical veracity and marketability, the latter would prevail.

This model of man acquires an unexpected support from the fact of giving a different image to computers. Computers became the embodiment of pure rationality, and this rationality became the primary characteristic of man, of being human. The next step, therefore, is to recognize this characteristic in computers as a mark of something more than being an inanimate object and electronic abacus. Man is a man because of the rational dimension. Man has rights because of being a rational being, and so should have computers, since they embody the pure essence of rationality proving in innumerable ways the efficiency, precision and elegance of what they are doing, proving that extracted and purified rational dimension works so much better. Hence, computers ought not to be denied the status which by nature belongs to men, the creatures which are distinguished from other creatures by their rational dimension. The problem of ascribing the same rights to computers as to humans is being posed, the debates about the fact that computers have personality are conducted, and the movement of computer rights is gaining momentum. This is no longer an isolated voice of technophiles, but opinions which may be found in many journals and books.

The problem of computers' personhood can be easily solved by resorting to the observation that language constantly changes. For instance, the concepts of universe and planet have a different meaning now than they had centuries ago. Thus, to consider a computer as a person only requires the definition of person to be extended or modified. However, this solution would hardly be satisfactory. What is needed is a reflection on the concept of a computer from the perspective of existing conceptual structures and a decision about whether or not a computer can be considered a person now or any time in the future. If a computer is a person, it should be treated as a person. It follows that the discussion becomes no longer purely academic, since it can have far-reaching social, ethical, and legal consequences.

The idea of computers' personhood can be refuted in three ways. First, in can be shown to what consequences it can lead, if we are in a position to accept them, if only in a distant future, and we will try to show some of them. Second, it can be shown that the supporters of this view make unsubstantiated or inconsistent transitions. Third, another theory can be presented that shows the impossibility of a computer being a person. Such a theory can be presented as a set of postulates that can never be proven (as is the case with postulates), but some arguments can be presented in their favor. To a certain extent, all three ways will be used in what follows.

A distinction is made between a person and a human being. Although all persons we know are humans, there is a possibility persons exist who are not humans. A person is a creature characterized by the following features: rationality, self-consciousness, self-reflection, moral responsibility, and ability to communicate (Olen 1983, 233; Flanagan 1991, 61-64). Already this statement can become a bone of contention in two respects. First, all concepts used in this definition are chronically imprecise and have no generally accepted definitions in their own right. What constitutes rationality? What is a distinctive feature of consciousness?   Second, not all would agree on this particular list of a person's characteristics. Is ability to communicate necessary? Is moral responsibility needed?   Being aware of these drawbacks, this definition is assumed here as sufficiently describing personhood. Equipped with this understanding of the concept of person we shall proceed to the discussion of the view that computers as more than mere tools.

"If being a conscious being with certain psychological properties - writes Chris Fields - is sufficient for inclusion in society, then a machine with those properties deserves inclusion as much as a human with the same properties," (Fields 1987, 5). This implies that if such machines do not exist yet, they will be built: conscious machines equipped with certain psychological traits. Thus, regardless of how little is known about the nature of consciousness and any psychological properties, it is assumed in this statement that it is just a matter of technological sophistication to construct machines endowed with them. It also seems that terminology, being used with reference to machines and computer systems, encourages such a treatment of computers. For instance, computers should be treated as members of society because "they are intelligent entities with which humans interact socially."

First, the claim that computers are intelligent entities is not obvious. Many computer programs behave in a way we would call intelligent; but it does not give them a status of intelligent beings. Computers have been programmed and they run according to a software system - written by humans (or by a program that has been written by humans). Because of the complexity of many systems, the programmer himself may not be able to predict all possible behaviors or responses of the machine. Nevertheless, the machine does not do anything original or inventive. The lady Ada's statement is still true:

a computer behaves intelligently, because it was programmed to behave in this manner, not because it *is* intelligent. Computers are as much (or as little) intelligent as typewriters, cars, or toasters, and ascribing any intelligence to them is simply a matter of metaphorical expression not to be taken literally.

Second, "humans interact socially" with computers, and thus computers are the members of the society. Humans, to be sure, interact with computers, but dubbing it a social interaction is unjustified. Interactive programs are those that request input from the user and generate some output. But in this sense, a vending machine is also an interactive device, since upon inserting a coin and pushing some buttons it ejects a requested product.

In this way, in fact, each machine is an interactive device, since it awaits some input and responds accordingly; for instance, a car turns to the left if the steering wheel is turned that way, and red lights on its panel expect a response from the car user. Would we consider it a social interaction?

Social interaction is an interaction between conscious members of society who are able to understand each other and participate in the same culture which enables that understanding and proper interpretation in the communication process. Such an interaction is very often performed indirectly through different media, it is not necessarily direct, for instance, when two people talk over the phone, or exchange letters. Such an interaction can be reduced to an action in one direction only; for instance, the author of a book communicates something to readers, but (usually) has no interaction with them. In the case of computer systems we have a similar situation: the user indirectly interacts with (or is acted upon by) the system designer, not with the computer. The computer is just a medium, a tool so efficiently screening the user from the programmer, that an impression arises that a true interaction with the computer is taking place. This realization stirred Joseph Weizenbaum to write his book after he saw how people interacting with his simple program *Eliza* (and its descendent, *Doctor*) "became emotionally involved with the computer and how unequivocally they anthropomorphized it" (1976, 6).

On the other hand, it may be claimed that Weizenbaum did not realize the full meaning of his program. He did not perceive abilities emerging from what he put in this program, nor did he properly appreciate extra-tool ingredients implanted in it. In effect, that is what

Fields says about expert systems designers who regard these systems as mere application tools. Expert systems not only assist people, but do all tasks; their solutions "are often accepted as correct;" and they often outperform human experts. Therefore, "our acceptance of some expert systems as *true* experts in their fields is *prima facie* inconsistent with the insistence that these machines are mere tools. One cannot, without stretching the definition, respect the judgment of a tool" (Fields 1987, 11). However, cars almost always outperform people when they try to race against them, table saws always outperform people when it comes to cutting wood, and hammers are irreplaceable in driving nails into wood. Moreover, if a compass shows north, and an odometer indicates distance, we respect this judgment, without seeing them as persons. They were constructed as experts in their domains, but it does not automatically imbue them with personhood. They are tools because they perform better than humans and they were built exactly for that purpose. It is hard to see that, as a side-effect, personality was implanted in them as well. The situation changes only quantitatively when including computers in the picture. We are impressed by their speed and precision, but they are not results of the inner life of computers. We are impressed by the speed of a bullet and precision of a micrometer without elevating them above the level of mere tools.

The problem of computer personhood can be approached on the social plane from a different angle. "What matters most is whether or not we are prepared to accept a particular entity as one of ourselves. What is critical is the attitude of the rest of the society"; "the key aspects of what makes a being a person are not the nature of the inner processes" (Dolby 1989, 324,335). In this way, psychological aspects of a person are declared irrelevant, and the problem of a person having some internal qualities is made a subject of a social agreement, or even a social *fiat*: a creature is a person since it has been announced by the society to be such and treated as a person by applying to it moral and legal codes. However, does a person cease to be a person if declared not to be one? Are slaves not persons only because the class of lords strips them of this honor? If the slaves recognize themselves as persons, whose opinion would prevail, slaves' or lords'?

On the other hand, there is a possibility of declaring a person something that obviously is not a person. Children treat their teddy bears and Barbie dolls as friends, converse with them, give them treats, hear their advice, protect them, and heal if necessary. Do these toys become persons only because they are treated as members of children's society? Idolatry and animism are other examples of this practice.

In both cases we are certainly hesitant to agree upon the status of persons ascribed to toys and idols. Relinquishing the problem of being or not being a person to a decision of society does not solve it. The society of old declared Sun to be a planet of Earth, but in spite of a longstanding and general agreement and in spite of "the attitude of the society" Sun did not change its nature and Earth always revolved around it. The natural scientists did not rely on this agreement but analyzed "the nature of the inner processes" of the Solar system.

We are facing the same problem in the case of persons: relying on societal decision does not suffice. Persons are social beings and an immersion in social life is crucial in their development as persons. Moreover, it is this participation that helps them to become persons that is important, not the attitude of society. Becoming a person is a result of intersection of inner processes and social influences. Therefore cutting off the former leads at best to partial explanations based upon an impoverished image of the proverbial black box. Discarding a psychological dimension leads to filling up the void by an arbitrary decision of society. Thus, if "people are prepared to treat it [a robot] as a person ... they will also be prepared to attribute to it whatever inner qualities they believe a person must have" (Dolby 1989, 335). The problem is, therefore, solved by attributing some properties to computers, thereby including them in the category of persons. Inner processes, if any, do not require any analysis. They are simply attributed by communal decision to any object society wishes to be endowed with personal features.

The position of computer personhood can be, however, defended on epistemological grounds, not just by reference to agreement. "What we discover within ourselves is what society induces us to find, [which] leaves very little for our inner natures that all persons have but robots cannot have"; "we tend to see within ourselves what our culture leads us to expect to see" (Dolby 1989, 328,335). An example here is that medical students may imagine that they suffer from every second illness they learn about.

Through the socialization process man grows into the web of categories, theories, superstitions, and ideas that are held by other people. He sees the world through the perspective of these categories and only a painfully long process can lead to discoveries that, for future generations, are trivial and obvious, but, for his contemporaries, they are outlandish and unacceptable. Breaking out from commonly held categories, whether scientific, religious, aesthetic, or common sense, is not an easy matter. It took centuries to discover that the Earth is round, stars are spread in an immensely large universe, monsters do not swallow ships, dirt does not generate worms, and the brain is needed for man to think. As members of society we organize the world, including our inner world, using these categories and applying them in an attempt to explain different phenomena. It can be safely assumed that we see, in the sense of sensory perception, in the same way as centuries ago, but we explain what we see differently and we structure the world in a different way. Today, burning wood looks the same as it did in the eyes of Georg E. Stahl at the beginning of the 18th century, but for him phlogiston was necessary as an element of an object to allow combustion, which was released when burning it. For us an element from the outside, that is, oxygen, is necessary to make the combustion process possible. Hence, it can be said that contemporary science induces us, or leads us, to see, if only with a mental eye, oxygen to be one of the elements of this process.

Cognitive processes do not take place in a social vacuum. Each cognitive subject uses mainly current categories and methods, perhaps without full awareness. The progress of science consists in breaking out from the current web of beliefs, from currently functioning "social contract" in the domain of epistemology, from paradigm of the day into a novel conceptual structure leading to different organization of the cognitive material. Although "we tend to see within ourselves what our culture leads us to expect to see", we have to realize that this seeing can be patently distorted. This has to be kept in mind in case of characteristics that do not lend themselves to an easy operationalization, such as our inner life, awareness, intentionality, etc.

Our culture has a tendency to remove everything that stands in the way of progress measured in terms of efficiency. This includes also the realm of human personality. Since philosophers for centuries have struggled with the problem of personality with no generally acceptable results, then a way of solution seems to be eradication of all elusive concepts by declaring them non-existent or at least irrelevant. Because all this progress, that is scientific and technological progress, is taking place, thanks to intellectual abilities of man, then what cannot be included in intelligence is pronounced not truly human and having little bearing on the problem of personhood.

Moreover, some intellectual abilities of man have been transferred to the computer, frequently with incredible results. The computer seems to be an embodiment of what constitutes the core of humanness, the intelligence, reaching in many areas an efficiency unattainable for man. If intelligence purified from all admixtures is given a chance, it truly prevails. Therefore, what is the use of other elements considered to be human? A creature becomes a true person if it is guided by intelligence alone; emotions, beliefs, etc. can only distort a smooth course of affairs.

Thus, computers made us finally aware, where is our true humanity, and what constitutes our personhood. Computers simply cannot be denied a status of person having themselves the essential part of personality.

The personality of the computer still seems to lack some fullness. There are two avenues open to alleviate this problem: either remove from human personality elements that computers do not possess, or implant in computers psychological features they do not have. In fact, both avenues can be chosen at the same time. The process of enrichment of computer personality can go hand-in-hand with the process of belittling the extent of human personhood. If a certain element is ill-understood or ill-operationalized (i.e., awareness, intentionality, or conscience), it is pronounced negligible, but if it can be expressed in computational terms, it is retained. These processes truly leave "very little for our inner natures that all persons have but robots cannot have".

Assuming for the sake of discussion that computers should be considered to be persons, would we be ready to accept everything that such a statement leads to? Should not computers, as persons, be treated differently than plain inanimate objects? If rationality is to constitute a distinctive mark of humanity, then such tools share a very important feature with humans. Should they be treated as humans, or as machines, despite the fact that they were produced by man? This indicates that assigning rationality, and consequently personhood, to machines may have an impact upon a proper attitude toward them.

Computers, as members of society, would deserve equal treatment with other members. For instance, they should not be hurt. But how can one hurt a computer? There are some attempts to assign feelings to computers, in particular, pain. This is irrelevant, says Fields, since computers can be hurt on a purely intellectual level: "computers can be hurt by thwarting their goals" (Fields 1987, 18). Assuming that they are able to have their own goals, how can they perceive them being thwarted? If they can make a comparison between their goals and reality, would we have the right to hurt them by disregarding their judgment and relying upon our own? How can we balance hurt inflicted upon computers with a necessity of acting against their judgment? Would we be forced to assent to their goals out of concern that they may be hurt?

But in this way everything can be hurt. If goal is defined as a tendency to reach a certain state, then each object can be hurt. For instance, if we do not allow an apple to fall, its goal being to reach the floor, we are hurting it. If we disregard readings of a thermostat, we are hurting it as well (it should be remembered that according to John McCarthy, a thermostat has beliefs and goals). Therefore, by using a certain definition of goal, man makes impossible his existence in this world, since virtually every animate and inanimate object can be hurt.

The idea of a hurt computer can be defended on the grounds that it is intelligent, whereas an apple and a thermostat are not, thereby restricting the population of objects that can be hurt. Everything now hinges upon the definition of intelligence. A computer's intelligence can be easily dismissed by stating that it is not a computer that behaves intelligently when solving problems, but a man who uses it and runs his program on it. Thus, the fact that the program was written by a human and put in the computer would clarify everything. However, it can be responded that for a believer, man is an intelligent being, although he has been endowed with intelligence by God who breathed into him spirit that can reason. Man can develop his faculties in the course of life, but the defender of computers can answer that the computer can do the same by using evaluation functions, weights, dynamically adjustable parameters, or any other device discussed in the area of machine learning. If one tries to show human superiority over computers by recourse to the fact that humans are living beings and computers are not, then this view can be rebutted by the statement that carbon chauvinism is assumed here, and life, theoretically, can be based on silicon also, not only on carbon. This view is espoused by Geoff Simons who sees in computers the next stage in the evolution of life, by Maureen Caudill for whom they are evolving into our successors, or by Hans Moravec, for whom the future society is a society of computers.

Computers are to deserve not only ethical consideration, but also legal. Since it is impossible to give a convincing criterion singling out humans from the rest of the world and since in many instances computers behave more intelligently than humans, computers will also acquire the status of legal personality, which, Marshall Willick says, is just a matter of time. At that moment, only sheer prejudice will force people to look for such a criterion (Willick 1985, cf. Narayanan, Perrott 1984). Thus, "civil rights for machines ... may soon be current" (Fields 1987, 19).

Granting them civil rights can make the problem of updating a program rather difficult. What would constitute computer personality and its uniqueness? Would it be just hardware? In this case, a computer would have as much personality as a light bulb, both being made out of metal and silicon. Or would it be software? Then one should not tamper at will with the crucial part of a being's personality. Therefore, would a change of computer program require a court's permission, or just the computer's consent? Also, production of computers would raise moral issues, since if they can be considered "a genuine life form in its own right which can think" then, as suggested, discontinuation of production and throwing away the partly finished robot "would represent a new sort of abortion and would have to be opposed" (Anderson 1989, 166). In any case we might expect "some machines that it would be prudent to treat as ends only, and never as means" (Caws 1988, 307).

As persons, computers will enjoy full legal protection in the light of the law, and as intelligent beings, they will be also held responsible for their deeds: for misjudgments, miscarried actions, and accidents. The law is to protect them from being hurt, but it also is to impose on them punishment for misdeeds. The question remains what exactly such a punishment should be: incarceration, turning them off for some time, putting them in a room with no air conditioning, assigning them run-of-the-mill tasks, reducing their RAM?

There are other problems with computer's personhood, as indicated by Michael LaChat. When assigning rights to computers one may turn to the United Nations' 1948 Declaration of Human Rights. Article Four requires that nobody can be held in servitude. "Isn't this the very purpose of robotics?" asks LaChat (1986, 287). Article Thirteen gives persons the right to movement. Therefore, should all computer contrivances be mobile? Would wheels be sufficient, or is some form of limb needed? Article Sixteen gives the right to marry and found a family. Maybe computers can be treated differently, since, as von Neumann has shown, automata can self-replicate. If such automata assume a physical form, they may need no family to maintain the existence of the species.

Machines are widely deemed to be intelligent, and because of this, they may also be endowed with personhood. However, the ethical and legal dimensions of this issue have seldom been mentioned (cf. Boden 1977, 496). This may mean two things.

First, the problem of ascribing thinking abilities to machines is not really treated seriously by those who do it and no one, in real life, would treat such a machine as deserving human treatment. Second, it may be a sign of a very low treatment of ethical issues, in particular, in scientific and technological discussions. Man is reduced to the rational dimension, pruned from ethical, religious, aesthetic, etc. aspects. Therefore, bringing up ethical issues in the case of thinking machines is irrelevant, or negligible. The fact that a machine thinks is to solve all possible ethical problems.

Probably it is both: we may be inclined to believe that machine thinking is not quite full-fledged cognitive activity. Therefore, such a machine does not acquire the status of a fully rational being, whereby it is still a tool and should be treated accordingly. Also, ethical issues are somewhat waning in this age of technology, having much lesser status than scientific and technological issues. They may be dismissed all too easily and suppressed as having little importance.

The view of considering a computer as a person can be defended as an 'as-if' view, or as merely 'an intentional stance.' From this perspective, the computer is regarded as a person solely from the pragmatic standpoint, without taking any firm stand on whether or not it really is a person. Like functionalism in psychology, this view leaves ontological questions aside and looks at computer personhood as a useful tool, or a metaphor, to explain computer's behavior and human-computer interaction. Lars E. Janlert lists twelve implications of the person view (1987, 326). Some of them seem to be displaced, since they are also consequences of the tool view, for example, the fact that computer is cast in one piece, is present, and is not self-sufficient. Some implications are specific to this view - such as being social, ethical, intentional, etc. - but these consequences can hardly be viewed as desirable. Quite the opposite.

Viewing computers as tools would not allow us to judge computers the same way as persons, since in this way "they are neither good or evil" and we would only be able to say that "they are functioning," which does not give a fair picture of computers. It is, however, puzzling why such a statement is insufficient. Computers are machines, they are tools, and hence they are simply functioning. They can malfunction, but it does not mean they are morally responsible for it. It is simply misleading if this responsibility for erroneous output is placed on computers. In thius way, the blame is attributed to errors in programs, to computer architecture, and the like, although software and hardware designers should be held responsible.

Also, the tool view is blamed for making people afraid of using computers because of their complexity and intricacy. Therefore, using and controlling computers should be much easier if the person view is used, since "we are all experts on persons" being persons ourselves (p. 332). If we assume that when using computers we are interacting with a person then the process of operating computer should become less strenuous and more efficient.

It is, however, incomprehensible how the person view would make operating computers easier. Such a view gives an erroneous impression that computers can do pretty much the same as persons, and only cause more anxiety and disenchantment when computer behavior does not measure up to what is expected of persons. The person view causes an opposite effect from what is intended. For example, oftentimes designers of user interfaces are tempted to personalize computer by having computers print messages in the first person, such as "Hi, I am Fred, I will tutor you in solving quadratic equations." But, as observed by Ben Shneiderman, the designers follow here "a primitive urge", and messages of this type simply "deceive, mislead, and confuse users" by creating an erroneous model of computer-user interaction, by creating a misleading impression about computer capabilities, and by producing anxiety. The person view, instead of alleviating anxiety, increases it. Therefore, "presenting the computer through specific functions it offers may be a stronger stimulus to user acceptance that the fantasy that the computer is a friend" (1987, 322-3).

Studies indicate that users prefer depersonalized, hence more adequate, treatment of computers. For example, positive reinforcement messages (eg., *You are doing great*) were much less preferable than simple numerical statements (eg., *7 positive answers out of 15 (47%)*) and they did not improve learning whereas mechanistic messages did. Also, personalized messages were seen as less honest than impersonal messages (eg., *I'm ready for a file name* instead of *Enter file name*) and boring in the long run (Spiliotopulos, Schackel 1981; Shneiderman 1987, 323-5).

The person view, therefore, gives no advantage to the computer user as these results indicate. But maybe these results stem from an inadequate comprehension of the computer nature? If this is the case, the nature should be unearthed, and this may be possible by viewing computers as persons. The person view is to help the user to see "what computer systems *are*," since computers viewed in this way have goals, and realization of this fact "spurs him [the user] to try to discover or reconstruct these goals" (Janlert 1987, 331). However, this does not lead the user to discovering what computers are, since it has already been assumed that they *are* persons and only details have to be filled in by searching for the computer's goals, intentions, points of view, insights, etc. In this way, people are misleading themselves by seeing goals where they are not, by ascribing to computers characteristics they do not possess, and by trying erroneously to detect in computers properties they have already implanted in them (either by wishful thinking or by programs). This hardly can be seen a "profitable view," since it is difficult to see who really profits from it. Hence, 'intentional stance' is misleading and may lead very easily to consequences analyzed in this chapter: to discussion of computer responsibility, computer rights, computer mental life, and the like. The view is misleading in that it paints computers in unrealistic colors and equalizes machines with humans. Whatever reasons for such claims, they can hardly be accepted.

References

Anderson David 1989, *Artificial intelligence and intelligent systems: The implications*, Chichester: Horwood.

Boden Margaret A. 1977, *Artificial intelligence and natural man*, New York: Basic Books.

Caws Peter 1988, Subjectivity in the machine, *Journal for the Theory of Social Behavior* 18, 291-308.

Dolby R.G.A. 1989, The possibility of computers becoming persons, *Social Epistemology* 3, 321-36.

Drozdek Adam 1990, Programmabilism: A new reductionism, *Epistemologia* 13, 235-50.

Drozdek Adam 1994, Awe and arrogance in science, *The Midwest Quarterly* 35, 136-50.

Fields Chris 1987, Human computer interaction: a critical synthesis, *Social Epistemology* 1, 5-25.

Flanagan Owen 1991, *Varieties of moral personality*, Cambridge: Harvard University Press.

Janlert Lars E. 1987, The computer as a person, *Journal for the Theory of Social Behavior* 17, 321-41.

LaChat Michael 1986, Artificial intelligence and ethics: An exercise in the moral imagination, *The AI Magazine* 7, 70-9.

Minsky Marvin 1988, Introduction, in M. Minsky (ed.), *Semantic information processing*, Cambridge: The MIT Press.

Narayanan Ajit, Perrott D. 1984, Can computers have legal rights?, in Yazdani M., Narayanan A. (eds.), *Artificial intelligence: Human effects*, Chichester: Horwood, 52-61.

Newell Allen, Shaw J. Cliff, Simon Herbert A. 1957, Empirical explorations with the Logic Theory Machine: A case study in heuristics, in: Feigenbaum E.A., Feldman J. (eds.), *Computers and thought*, New York: McGraw-Hill 1963.

Newell Allen, Simon Herbert A. 1958, Heuristic problem solving: The next advance in operations research, *Operation Research* 6, 3-22.

Olen Jeffrey 1983, *Persons and the world*, New York: Random House.

Servan-Schreiber Jean-Jeacques 1979, *The American challenge*, New York: Atheneum.

Shneiderman Ben 1987, *Designing the user interface*, Reading: Addison-Wesley. Spiliotopulos V., Schackel B. 1981, Towards a computer interview acceptable to the naive user, *International Journal of Man-Machine Studies* 14, 77-90.

Torrance Steve 1986, Ethics, mind and artifice, in: Gill K.S. (ed.), *Artificial intelligence for society*, Chichester: Wiley, 55-72.

Weizenbaum Joseph 1976, *Computer power and human reason*, New York: Freeman.

Wilkes Kathleen V. 1984, Is consciousness necessary? *British Journal for the Philosophy of Science* 35, 223-43.

Willick S. Marshall 1985, Constitutional law and artificial intelligence, *Proceedings of the Ninth IJCAI*, Los Angeles, 1271-3.

Wooldridge Dean E. 1969, Can mechanical man find goodness, truth and beauty? *Psychology Today* 2, 11, 26, 28-9, 64.

CHAPTER 2

MORAL DIMENSION OF MAN
AND
ANIMAL RIGHTS

The problem of an ethically and legally proper treatment of computers is not without precedent, as we can find a parallel in the debate of animal rights. Animals have always been used by humans for a variety of purposes. They have been used for food, clothing, pets, entertainment (e.g., races), and experimentation. Some animals have enjoyed a privileged status, such as cows in India or animals constituting totems of some clans; however, in most cases they are not held in such a high esteem. Is this right? Should we treat animals on an equal footing with tools? Can they be used for any purpose imaginable and in any way we please? Do we have a moral responsibility toward animals as we do toward people? These are the questions that not only give rise to theoretical discussions, but also lead to very concrete steps undertaken by various organizations such as People for the Ethical Treatment of Animals, or the Animal Liberation Front. These organizations are fighting for animal rights and for decent treatment of animals both in their natural settings and in laboratories. The use of animals for medical research has become an especially sensitive and personal issue. Can we inflict pain upon animals when testing methods and medicaments that may save human lives? The answers are sharply divided and each side adamantly defends its own view.

One particular philosophy behind the struggle for animal rights relies on evolutionism and draws it to a seemingly inevitable conclusion: animals do have rights. Man is an animal, says Richard Ryder, who "arrogantly exaggerates his uniqueness;" but along with accepting our "biological relationship with other animals" we should take a next step and "acknowledge a moral relationship" and treat animals as our relatives, thereby stopping all research on them. The next step would be a rigid vegetarianism; but what to do about pest control? Can it be considered cruel to annihilate those with whom we have a moral relationship? Also, what to do about scientific experiments on animals? If they would not be banned, then how do we determine which experiments are unnecessary? Drawing from Peter Singer's ideas, Edwin Hettinger says that we should try to answer the question: "Would the investigator still think the experiment justifiable if it were performed on a severely retarded human at a comparable psychological level as the animal? If not, then the experiment should not be conducted" (p. 126). Only "arbitrary preference for members of our own species," the preference abhorred by Hettinger, would allow us to conduct the experiment. Thus, there should be no difference in treatment of a rat, an ant, a lizard, or a severely retarded human whose intelligence does not exceed the IQ of these animals. However, animals are not humans and they should not be treated on an equal footing with them. If no difference is seen between these two worlds then, to be sure, we can be appalled with Peter Singer who says that "even the most profoundly retarded human being is entitled to the respect and moral consideration that we properly deny to the most intelligent dog" (1991, 61) and after Richard Dawkins that "a human fetus, with no more human feeling than ameba, enjoys a reverence and legal protection far in excess of those granted to an adult chimpanzee" (Dawkins 1976, 10-11).

In discussions concerning the possibility of animals having rights, the same strategy is always used: animal rightists try to find a property common to both humans and animals, thereby proving that animals should have the same rights as humans. The opponents try to find a property that sufficiently distinguishes humans from animals, showing that animals cannot share rights possessed by humans. The latter are mainly interested in the old philosophical problem of who we are, who is man, what is human in man. In the face of new discoveries and advancements of science and technology, they try to reassess the position of man toward the rest of universe, which may serve a purpose of proving the rights to having dominion over all things, but certainly they attempt to find what is specific in man. Animal rightists, environmental ethicists, or computer personalists, on the other hand, attempt to dissolve the specificity of man by showing that all human properties are, to some extent, shared by other beings or objects, that there is nothing particularly special about man and all claims to the special position of man in the universe are exaggerated, if not an outright expression of speciesism comparable only to racism or chauvinism. This equalizing procedure can be performed out of reverence to nature tinted with misanthropy, which leads to bringing man down to the level of other beings or, out of reverence to man, by lifting up other beings to the level of man (the way Hinduism does by including them in the cycle of reincarnation). Whatever the motive, the procedure is the same: finding a property that proves that there is no significant difference between man, animals, and, possibly, inanimate objects. One of these properties very widely discussed is the posse-ssion of interests.

The problem can be established just by definition. Leonard Nelson having "defined animals as carriers of interests" (1956, 116) and having assumed that interest carriers have moral rights, immediately solves the problem. Thus animals have such rights, and since moral right holders are persons, animals also qualify as persons. There is also a possibility that plants and stones have interests ("for all we know, the stone may have an interest in being trampled by us, or a cabbage in being eaten", p. 140), which populates the world with persons almost without limits. However, taking the definition of animals literally, we can argue that the animal world is much smaller than usually thought, since only those beings qualify as animals which have interests. Hence, if it can be proven that worms and crustaceans have no interests, it would mean that neither are they animals. Providing the possibility of stones having interests, there is no need to exclude worms from the kingdom of animals. Besides, Nelson gives no analysis of the concept of interest, therefore it would be hardly operational in deciding whether worms can have rights.

Other thinkers are more specific in their discussion of interests. "An interest ... presupposes at least rudimentary cognitive equipment. Interests are compounded out of desires and aims, both of which presuppose something like belief, or cognitive awareness" (Feinberg 1974, 52). It is clear that "having interests" means here "being interested in," or "express an interest in something." Tom Regan remarks that "having interests" can also mean to be good for someone, to be in someone's interest and imposes this meaning on Feinberg's understanding of having rights which brings him to the conclusion that trees, stones and cars can also have rights (Regan 1982, 170, 178). Therefore, as Regan claims, it is in his car's interest to put antifreeze in it in the winter, and only indirectly it is in our interest. It is clearly playing on words and in a similar tone it can be claimed that mother Earth is a loving being, since loving people attract other people and the Earth attracts all objects that fall toward its surface.

Feinberg says that "a person without interests is a being that is incapable of being harmed or benefited, having no good or 'sake' of its own" (p. 51). Therefore, beings with no good of their own have no rights. Regan disagrees with that and says "that we can make sense of the idea that beings who cannot be interested in things nevertheless can have a good of their own, and that we can, therefore, literally speak of what will benefit or harm them" (p. 175). He makes a distinction between a value assigned by us to objects and their having a good of their own. Thus, we value a car because it has some goodness independently of whether we see any value in it or not. "A good car does not lose its goodness if we lose our interest in it" (p. 178). What Regan seems to do here is replacing an old philosophical concept of essence by the concept of goodness. A car possesses certain properties that we value (or do not value) and these properties do not vanish whether we perceive them or not. It is this 'carness' of car that allows us to use it as a car. If these properties change, a car becomes a non-car with an essence of its own, or with a good of its own. The name of these properties could be irrelevant, unless it serves proving certain point which would be difficult to use with other words. By exchanging 'good' for 'essence' Regan is able to show that inanimate objects have some good in them, *ergo* can be harmed, *ergo* should have rights (of moral or legal nature) and should be treated accordingly. The concept of essence would hardly entail this line of reasoning.

Animals are protected by law from cruel treatment, from extinction and the like, but, Regan says, it is not satisfactory, since what is also important is the the basis of such legislation (p. 160). We are unjustified in elevating ourselves above animals, and therefore, in denying animals laws we ascribe only to ourselves. Animals should have an equal moral standing with us, "not just because we are like animals, but because ... we are animals" (p. 159).

From the evolutionary standpoint this statement is obvious, but should it mean equal treatment of animals and humans? Humans are also clouds of atoms, but probably no one would argue for the equal treatment of all clouds of atoms. We are clouds of atoms when considered from the perspective of quantum physics, but it does not constitute our humanity, and, in this sense, is irrelevant. Therefore, it is justified to say that no, we are not merely a cloud of atoms. Similarly, bringing up an animal ancestry of the human species may be of little relevance in analyzing the humanity of man and achievements of the human race. In this sense, we are not animals, or rather, history of man owes its course to what surpasses the animal in man. Regan admits that there are some "significant differences between the human and the other animal species," but he hastens to add that these differences "can no longer be supposed to be as great as was once widely believed"; these "significant differences", in fact, become so insignificant that animals should acquire equal treatment along with humans.

Another property used in discussions of animal rights is sentience. This property is not unrelated to interests, since "the only condition that would need to be met in order to have a moral right is that one has interests, and this condition is met by all sentient beings" (R. Godlovitch 1972, 158).

One possible understanding of sentiency is given by Singer, who bases the claim of animal rights on this characteristic. He says that sentiency is to mean the "possession of a nervous system remarkably similar to our own" (1976, 12, 185). Even this purely biological justification of moral rights possession causes a problem in determining any remarkable similarity of various nervous systems with ours. Would all mammals belong under that category? Is nervous system of birds and fish sufficiently remarkably similar to human's? Singer realizes these difficulties, and later somewhat changes his substantiation of animal rights by stating that "the only acceptable limit to our moral concern is the point at which there is no awareness of pain or pleasure, no conscious preference, and hence no capacity to experience suffering or happiness." However, the same problem reappears, since it is not obvious where to draw the line between beings able to suffer and not able: "I can only plead agnosticism," admits Singer (1991, 64).

Other specifications define sentiency as a possibility of having sense perception. It may not be clear whether one sense would be sufficient, or a specific number of senses has to be possessed to become a right holder. Some animals do not see, some do not hear. On the other hand, most plants can be seen as feeling entities, since they react to the changing position of the Sun, and some of them, carnivorous plants for example, feed on insects reacting in a certain way when an insect touches them. Is then, the Venus fly trap a moral rights holder because it is an entomophagite and carrot is not, since it draws its food from the soil?

Whichever meaning is assumed there are some problems which Robert Wright faces, who sees in sentience "a condition for the possession of high moral status", and not in reason or self-consciousness, but in this way he is unable to argue why killing 100 baboons to save a human life is not the same as killing a human to save 100 baboons "unless you can create a moral ratchet called 'human rights'" (1991, 48).

One of the arguments against animal rights is that "only persons can have legal rights" and animals are not persons. But, Regan states, "corporations are persons within the law, as are ships." Therefore, it may be only a matter of time for animals to become persons (1982, 152-3). However, Regan seems to treat intercheangably two meanings of 'person': legal person, and person proper. Corporations and ships are named legal personalities by law and for the sake of legal statements thereby acquiring no status as a person in the same sense that humans are persons. Corporations do not cease to be tools through which humans act and interact. They do not become living and thinking entities endowed with a soul of their own in some Durkheimian sense. Corporations' personality is a legal creation serving the purpose of enacting laws prescribing and prohibiting some behavior of people toward people. A corporation is just a medium through which people organize their interaction among themselves, it exists through people and for people (oftentimes also against people) and in this sense has no separate standing and no independent personality. Corporations remain as much persons as ships, stones, and trees. Legal terminology does not create real personhood.

Of course, law can name animals persons (legal personalities), but their personhood would remain as impersonal as that of corporations. Calling animals persons will not endow them in the properties we would like to see in persons. Naming animals persons will liquidate the gap separating animals from humans only verbally. Thus, laws protecting animals should be enacted on account of humans and not animals. It is by serving and enhancing what is human in humans that animals can be protected best. For instance, hunting for pleasure is detestable because it serves the cruel and animal part of ourselves and therefore is, at least, undesirable. Banning it would be in interest of our human part. Therefore, stressing in this instance that we are animals to protect them has an opposite effect: the animal in us is simply against such a protection. The human part of us strives for development and protection of this part, which has a saving effect of protecting non-humans as well. Recoursing to animal personhood would be just a verbal act, having as little impact on our morality as rather frequently the law has.

When arguing in favor of animal rights, the argument of the rights of children and the mentally impaired is invariably used. The argument has two forms. The first is an argument to the effect that a criterion that excludes animals from right holding beings also excludes children and the mentally impaired. As Andrew Linzey writes "if we accord moral rights on the basis of rationality" then "logically, accepting this criterion, they must have no, or diminished, moral rights" (Linzey 1976, 24). That is, the argument from rationality is too strong, since, along with animals, it would exclude children and mentally impaired from among beings that should have rights. What this form of argument implies is that the criterion used in it is unacceptable and, also, that animals should be included in the right holder class, otherwise children and the mentally impaired may be left without any rights. One possible solution is simply calling some humans non-humans. At one time Regan excluded irreversibly comatose people from the class of people, but later he called this argument foolishness and plain falsehood (p. 128).

The second form of argument that refers to rights of children and the mentally impaired uses some criterion that guarantees inclusion of these humans along with animals in the class of right holders. That is, since we ascribe moral rights to humans devoid of certain characteristics deemed specifically human, so we should also assign them to animals. For instance, the argument from sentience is inclusive enough to embrace children along with animals among beings having rights. The problem is, then, in finding a criterion that would decide what are beings that can be ascribed rights and what beings have no rights.

There were many different criteria proposed throughout the history of philosophy, for instance, free will, having interests, autonomy, consciousness, possession of soul, of language, of culture or of intentionality, participation in social life and the like, and all of these failed to resolve properly some borderline cases, which, on other grounds, seemed to require a different solution. Some of these criteria were intended to distinguish humans from other animate and inanimate beings. Some of them were designated as arguments for creating a class of beings that may be ascribed rights. This class would include also humans, and in this case the specificity of humans would be of lesser importance. Moreover, none of these arguments were undisputable; there have always been some quoted cases proving that they are too restrictive (e.g., rationality), or too inclusive (e.g., sentience). There are at least two avenues open to remedy this difficulty. One way is simply replacing a criterion by one more adequately suiting needs of a certain theory. Another way is to relax or to strengthen conditions of a certain criterion to include or exclude certain beings from a class under discussion. For instance, a sentiency criterion can be stretched to prove that plants should have moral rights by claiming that plants are sentient (Tomkins, Bird 1973); on the other hand, to restrict these rights to humans only, sentience can be taken from animals and limited to humans alone (Harrison 1991)[1].

Sometimes this class cuts across humans and animals including only some of them. An example of this approach is Regan's criterion of inherent rights. This criterion "leaves it an open question, whether these [comatose] humans can or cannot, do or do not have basic moral rights" (p. 141). However, this criterion does not leave any doubt about possession of these rights by some animals. He also finds that solving this problem "is not essential" since the soundness of his argument does not hinge upon it (p. 142). Thus what counts to him is the soundness of his argument rather than ethical status of some humans.

A favorable interpretation of Regan's criterion is that it, in fact, is a meta-criterion, since it does not specify what this criterion is, but only gives conditions it should meet to be accepted. It leaves an open question of moral rights, not only of comatose people, but all beings altogether (although Regan's treatment of his criterion does not prove it). Thus, according to the definition of an inherent value, a being x's having inherent value does not depend on other being's assessment of x; because of this, value x should not be treated as a means, and x should be treated with respect (p. 133). This specification gives only "noteworthy features" of an inherent value and by using just these features, one would be unable to say what an inherent value can possibly be. In his discussions, Regan refers to a possibility of a being

---

[1] Harrison uses Descartes' name in support of his thesis, which is a rather popular misconception of his views. Descartes was of the opinion that animals were machines, but this did not entail that they felt no pain. In one letter he wrote: "I should like to stress that ... I deny sensation to no animal in so far as it depends on a bodily organ" (quoted in Cottingham 1978, 557).

to improve its life, but it is not entailed by the noteworthy features. These features would simply indicate that there is something specific in each being which makes it what it is, regardless of our seeing it. But, as he indicates elsewhere, "inherent value ... must be discovered" being "an objective property" of an object (p. 199).

Although very similar in its ontological significance to the concept of essence, inherent value cannot be equated with it any more, since, first, "not everything in nature is inherently valuable" (p. 200), and all things have an essence. Second, Regan changes his mind and rejects the view that inherent value is being good of its kind as "completely muddled". Now, the property which would make inherent value different from essence is admiration: "the inherent value of a natural object is such that toward it the fitting attitude is one of admiring respect" (p. 200). Would it mean that what does not stir in us an admiring respect has no inherent value? Regan certainly would deny it but his attitude indicator is hopelessly subjective. Therefore, his inherent value criterion is reduced to the statement that certain entities (but which ones?) have something in them (but what?) on account of which they should possess certain rights. That much is simply common sense and can be stated without lengthy argumentation. We still remain in the cloud of metaethics, hoping for our sense of respect to resolve an issue. Regan realizes it and asks himself, "How, then, are we to settle these matters? I wish I knew. I am not even certain that they can be settled in a rationally coherent way" (p. 203). The conclusion is of little help to lead us from the darkness.

What I looked at before were arguments used by animal rightists, but it is interesting to notice that their argumentation is often very strongly laden with emotional rhetoric. "Arguments that defend the use of animals for food ... are about as worthless as those defending human slavery" (Harris 1972, 97); "Speciesism is just a moral mistake of the same sort as racism and sexism" (Hettinger 1991, 120). Such statements are not infrequent and can be explained by treatment of animals by humans, who used them at will killing for pleasure, treating them cruelly like mere objects. It is obvious that animals suffer and can feel pain, and this fact alone justifies a proper treatment. Welfare of animals should certainly be our concern and most laws aiming at protection of animals and their welfare is undoubtedly justified. What animal rightists want is to go one step further and proclaim animals to be persons deserving moral rights, human position, and dignity. They want to break (for whatever reason) "arrogantly exaggerated uniqueness" of man and place animals (let alone tress, rivers, etc.) on equal footing with man in an all important respect. They want humane treatment to be turned into human treatment, to put aside biological differences and create one platform of moral rights holders for all (animate) creation. It is said that "once we acknowledge life and sentiency in the other animals, we are bound to acknowledge what follows, the right to life, liberty and the pursuit of happiness" (Brophy 1991, 128). What follows is an introduction of vegetarianism for all - and here is a problem. Humans can be legislatively forced to become vegetarians, but how can we obviate the problem in the case of carnivores? Should we pursuit the happiness of lions or of antelopes, is the liberty of sharks more important or of tuna fish'? True, "in granting that someone has a moral right, one recognizes his liberty to pursue those interests that are compatible with like interests of others" (R. Godlovitch 1972, 158). But precisely in what sense the interest of an eagle can be compatible with an interest of a mouse, its prey? Therefore, it is often forgotten that the practical consequences of statements spelled out in the fervor of discussion go well beyond laboratories and slaughter houses from which animal rightists would like (very often, for a good reason) to save animals.

Therefore, it seems that in the fervor of discussion concerning animal rights somehow an element of common sense disappeared. Despite superficial similarities, there is a gap between the worlds of animals and men. Impressive as they may be, the achievements of the society of ants will never measure up to those of humans, and the language acquired by Washoe will never attain the level of the first grade student. Biologically, animals are relatives of man, but it does not mean that they are equal to men in all important respects. There-fore, if there is a problem, whose life to choose then the life of humans is always more precious than the life of animals, no matter how retarded the humans may be. The same common sense tells us that it is inappropriate to kick or kill animals for pleasure, cause them unnecessary pain, conduct redundant research involving suffering of animals, hunt for pleasure, etc. It is by no means an indication that there is only a negligible difference between animal and man. But where can this difference be found? How can we define the distinction between the realms of animals and men? The question is not new, and it has preoccupied philosophers for centuries. In fact, all sciences revolve around this problem - the problem of human nature: "It is obvious - David Hume wrote - that all sciences remain in a more or less clear relation to human nature".

In all these discussions on whether or not animals (and inanimate objects) have moral rights, always the same approach was applied: animals have these rights because according to such-and-such criterion they belong to the same class as humans and, since humans have moral rights, so they should not be denied to animals. The criteria, as we have seen, are of a different nature, but they more or less distinctly clustered animals and humans in one class. Virtually, there is no limit to the number of different criteria that would delimit this class, maybe even better and more decisively, but they would not be as suitable for the purpose of arguing for moral rights of those included in the class as criteria that are usually used. For example, the class can be created by making a simple list of beings included in it: the class of moral right holders is the class of humans-and-rabbits-and-cows-...-and-eagles. This would be an extremely cumbersome way of doing it, but rather precise. Another way would be by approaching the problem the way non-monotonic logics try to resolve: the class under discussion includes featherless-beings-with-flat-fingernails-and-if-they-have-no-fingernails -then-with-a-head-but-with-no-scales-..., a definition with no end and somewhat less precise. But, it would be unsatisfactory, since the point is not just in defining a class of beings, but to prove that the criterion used in this definition entails another property that the members of the class have to share only on account of using this criterion. Thus, just as the criterion of having four angles singles out objects that not accidentally but by this criterion have also four sides, so a criterion of sentiency or awareness or interests is to entail that the beings meeting this criterion are also moral right holders. The crux of the matter is that, in most cases, it is not the case. As Raymond G. Frey observes, the animals rightists "*implicitly assume* that having experiences or mental states is valuable in its own right" (1980, 46). That is, experiences or mental states are assumed to have a bearing on the discussion on having rights as though it were obvious. We can ask, then, what should we imply from the fact that we created a set of beings having mental states? Does it mean that all of them have the same moral rights? This is a transition which is invariably lacking and that is the most crucial. The rationale behind such criteria seems to be always the same: we want to prove that these beings have the same rights as those beings. Therefore, let us find a criterion that joins the classes of these beings in one thereby proving our point. Or as Frey phrases it, "in the

hands of animal rightists, a sentiency criterion ... seeks for the lowest
common denominator between men and 'higher' animals, endows ...
this denominator ... with moral significance ... But why should we seek
for the lowest common denominator? ... [I]f we do not ... animals are
unlikely to be included within the class of right-holders; but this is no
answer at all, to one who is unsure whether animals possess moral
rights and who requires an argument to quiet his doubts" (p. 51).

The solution put forth by Frey is to renounce an existence of moral
rights altogether. To him, "the history of ethics reveals that it is by no
means an easy task to show that human beings do possess moral
rights," therefore, "either moral rights are superfluous or we are not
yet in a position to affirm that there are any" (pp. 7,17). There are
some other philosophers also who have doubts about their existence
(Kleining 1978; Young 1978).

Another, not so drastic a solution would be to cease to look for a
more basic property on which morality can be founded. It requires
life, intelligence, awareness and possibly other properties also, but it
is new quantity not entailed by any other properties. It is the main
contention of this book that the most distinctive feature of man is the
moral dimension. This dimension is what primarily differentiates man
from other beings, and it serves a basis for discussion of moral rights.

Animal rightists adamantly argue in favor of equal moral rights for
animals as for humans. To be sure, they do not think moral rights are
superfluous. However, what is the status of these laws in the absence
of humans? Would it be justified to think about a possibility of
ascribing moral rights to animals before man appeared on the Earth?
Would it make sense to talk about dinosaurs having these rights? An
affirmative answer would sound very factitious.

Moral dimension, as stated, is a new quality which appears along
with man. Only with this dimension do moral rights appear. Man is
a repository of these rights; only with him and through him do moral
rights begin to exist in the world and radiate to the rest of the world.
Through man, the world becomes saturated with morality and only then
this world can be viewed and scrutinized in the moral perspective. As
men, we are endowed with this dimension and we pass it over to other
beings in form of moral recognition, which may be viewed as second-
hand moral rights. That is, animals by themselves have no moral
rights, and yet they can be treated in a morally acceptable (or unacceptable) way. Therefore, "it is nonsense to suppose, as many animal
rightists do, that the denial of moral rights to animals leaves them

utterly defenseless" (Frey 1980, 170).   Ascribing moral rights to animals would be only a verbal manipulation having as much weight as ascribing such rights to their toys by children.   We may pretend animals have these rights and it may even change our behavior toward them, but it would not change the fact that animals do not have such rights.

Man is not the same as an animal, and in spite of many animal characteristics man possesses, what is genuinely human in man goes beyond what can be found in animals.   Man prides himself with intelligence incomparable to what can be found among animals - and this is true.   But just intelligence would make us automata, and, paradoxically speaking, thoughtless intelligent robots.   The greatness of man was built by his intelligence, but intelligence by itself would not be able to lead man from caves.   Intelligence has to be guided by what is truly human; by morality, by consideration of what is good and evil. This consideration frequently dictates something which intelligence cannot accept, but it has to, since we are men.

Because of the moral dimension that constitutes us, we recognize moral rights of those who are not aware of that, but being human they are also by nature moral beings.   We recognize the rights of children, including unborn children.   There are heated debates from what week such rights can be ascribed and how many weeks a woman has in order to dispose of an unwanted fetus without infringing upon its rights. However, when a child is born, there are no doubts about its having such rights.   The reason for recognizing these rights is that a child, being human, is a bearer of morality within it despite the fact that it can become aware of this dimension only as an adolescent.   We recognize these rights also in the case of the mentally impaired, although they may be incapable all through their life tell right from wrong. This inability, however, does not reduce them to the animal level; they still are humans with defective intelligence, awareness, or moral judgment, but still humans and hence, beings possessing moral rights.   Their inability to act morally or intelligibly does not dispossess them from the rights that they bring to this world by being born into the human race. Analogically, these rights are not suspended for people who are sleeping, hypnotized, intoxicated, seriously ill, or who fainted, and their inaction (or action, as in sleepwalking) may indicate that they know nothing about right and wrong (or about anything, for that matter).   But in these situations, their rights become even more important than usually, since they are more defenseless, and at the mercy of others. In most cases these rights are recognized and respected. Therefore, we can only agree with Cherry who says that describing infants as potential persons and the senile as lapsed persons is grotesque. "Personhood is not a distinction to be won and lost but a condition bestowed by membership of a particular kind," hence if anencephalic neonates "are not persons it is because they are not human beings either" (Cherry 1989, 348).

The rights people have, and animals do not, consequently lead to valuing more human life than animal, which is, as Godlovitch dubbed it, the greater-value principle (p. 163). This principle may be a good test of how seriously animal rightists treat their thesis on equality of human and animal rights. The cofounder of People for the Ethical Treatment of Animals, Alex Pacheco, says: "You and I are equal to lobsters when it comes to being boiled alive. I don't mean I couldn't decide which one to throw in, myself or lobster" (quoted in Wright, 1991, 43). Which means, that when it comes to a crisis situation, human life would be preferred. Other rightists are not so generous. After introducing her greater-value principle, Roslind Godlovitch comments, it is "a guideline for crisis situations rather than a principle" and "the sense in which it does it is the same sense in which the welfare of a loved one would be valued more highly than a stranger's" (p. 164). That is, if we love our hamster, there is nothing improper in giving it a preference in a crisis situation over a stranger or a disliked neighbor. In a similar tone, Singer requires "the interests of every being affected by an action are to be ... given the same weight" (1976, 6), which expresses what Regan called an "uncompromising egalitarianism" (1991, 82). It also may lead to a practical difficulty in a life saving situation if the interests of animal have to be treated on a par with those of man. For how long should these interests be weighed in order to be sure that, after all, human life should be chosen? Or perhaps, it may turn out that interests of an animal may outweigh man's.

Only one observation in passing concerning the claim of existence of the morality among animals will be made. For instance, dolphins are altruistic toward wounded dolphins, animals protect their descendants, etc. These are undeniable facts, but it is a matter of on what basis they perform these acts. The same facts can change their significance depending on perspective from which they are viewed. Thus, animals can be seen as able to use language because of some experiments with Washoe or Koko. But would we call a tape recorder enabled with this ability because it, in a sense, can speak? Are animals, literally, intelligent because they can find food, recognize other animals, etc.? But in this way, a lamp that turns itself off after two hand claps is intelligent too. To return to our question, are animals altruistic in the same sense that humans are? Is a smoke detector altruistic when it warns people that the house may be on fire? The same outward characteristics can be a result of different causes in different beings and situations, the problem widely discussed in the methodology of science. Whether moral-like behavior of animals is attributed to animal morality largely depends on the theoretical framework in which it is analyzed. The contention of this book is that despite statements of ethologists or sociobiologists concerning morality in animals, this behavior is as much a result of instinct as their alleged possession of language or intelligence.

All these discussions revolved around animal rights. It is interesting, however, that the animal rights advocates do not discuss the problem of animal responsibility. They want just rights for them. Should animals, as the right holders, be punished for misconduct? The problem was addressed in antiquity in many laws. For instance, the Mosaic law required stoning an ox that killed someone, and it was not satisfied by sentencing its owner to death by stoning if he knew that the ox tried to gore before and did not take necessary precautions (Exodus 21:28-32). The goring ox was also a popular topic of the Middle East laws: in the Babylonian code of Hammurabi (secs. 250-2) and the laws of Eshnunna (secs. 53-5), the owner of a ox that gored a man to death had to pay a fine. Thus, Mosaic law was more severe than their predecessors and it also specifically required sentencing the ox involved in killing.

In ancient Greece, in MacDowell's words, "an animal found guilty of homicide was presumably put to death or driven out from Attica. An inanimate object, such as a stone or a tree which had fallen on a person and killed him, was cast beyond the frontiers of Attica." The word 'presumably' should be stressed here since actual trials of animals are very uncertain and Aristotle is the only classical source to explicitly mention the trial of animals, but it is nor certain that these cases were held in court (the Prytaneion) and if it was a trial conducted in the same manner as trial of men. MacDowell and Hyde (1916) do not analyze these problems. MacDowell is inclined to say that it was not a trial but a very meaningful ritual: "If someone was killed by an object, an animal, or an unknown person, it was desirable that the state take note of the manner of his death, and take any steps that were practicable to see that no one else died in the same way in the future. This court served some of the purposes of a modern coroner's court" (MacDowell 1978, 117-118).

With regard to other parts of the world it is enough to quote J.J. Finkelstein: "Societies of non-Western derivation and primitive peoples *did not and do not* attribute 'human' will or 'human' personality to animals or things, and *never* have tried them or punished them as they did human offenders" (1981, 64). That much about the view that "in primitive law, animals, and even plants and other inanimate objects are often treated in the same way as human beings and are, in particular, punished" (Kelsen 1945, 3), which is substantiated neither by historical nor ethnographic data. Therefore, examples concerning trials of animals can be found only in Western civilization.

In medieval Europe, animals were frequently subjects of procedures aimed at invoking supernatural power to influence people rather than animals. They were more of a religious character than legal trials, since animals (rats or vermin) were viewed as an instrument of divine punishment leading to repentance of people for sin that caused that predicament. For instance, in 1120 a bishop of Leon excommunicated the caterpillars ravaging his diocese, in 1565 "the grasshoppers ... were ordained to quit the territory [of the Arlesians], with a threat of anathemization from the altar" (quoted in Stone 1987, 65). Authentic trials began in thirteenth century and most of them involved pigs sentenced, for instance, for eating a baby.

Before we condemn the absurdity of this procedure, and before we pride ourselves that such "examples illustrate well the progress we have made in our ascent toward a more enlightened view of the world" (Regan 1982, 152), we should realize, as Christopher Stone comments, that "to our recent ancestors, it seemed just as mindless to prosecute corporations ... as it seems to us to prosecute oxen" (p. 65). But subjecting objects "to criminal process may be viewed as an effective although indirect way of modifying the behavior of the people best positioned to exercise control over instruments of harm. In the one situation we are sending a message to those who can tether or fence the oxen; in the other, to those who can steer the corporate bureaucracy" (p. 66).

In all cases when an animal was tried, animals were executed not because they were treated as responsible beings, and not primarily because they endangered people physically, but because they were damaging moral, ideological and religious foundation of society. The world was divided into three spheres - God, man, and the rest of nature - and animals that were lower in the hierarchy simply could not act upon humans, whose position in this hierarchy was higher. If they did, they could not be suffered any more and should be disposed of the same way that faulty devices are destroyed: because they could not be trusted and they exposed people to danger and damaged the ideological fabric of society. "The stoning of the offending ox - which in medieval Europe was appropriately translated into the burning or hanging of the offending animal - was, therefore, the only effective way of dealing with the *objective* danger to the universal order which its action had created" (Finkelstein 1981, 47).

In face of these findings, it is not so certain that these "examples illustrate well the progress we have made in our ascent toward a more enlightened view of the world." This treatment of animals, strange as it may be, does not prove senselessness of men of old; it points to their attitude toward position of humanity in the universe, which was debilitated by behavior of animals. The system of values which stood behind it was not altogether senseless, only acts were lacking reasonableness - from our perspective. People were lacking knowledge, which we have, but we are lacking an appropriate perspective toward the issue of man vs. animal. Knowledge suppressed the spiritual, and in particular, the moral dimension of man, so that what remains is scientific treatment of man and animals in the sense of seeing both as mere variants of a preceding life form. According to this view, nothing new appears on earth with emergence of man, the same animal world is evolving from one physical form to another. In this view, we are forced to see animals as our equals and to be certain that the progress of science and technology is a guarantor of "a more enlightened view of the world." In some respects it is, but if only little difference is perceived between the worlds of men and animals we can wonder where this enlightenment is to be found. If man is seen as only a biological being then these two worlds simply have to be treated in the same fashion. Regrettably, compassion and altruism which in many cases guide the animal rightists disappear in their theoretical discussions where interests of animals are defended mainly on biological grounds. The moral dimension never appears, and this dimension would be a proper key to humane treatment of animals. Blurring differences between animals and men will not lead to such a treatment. Only emphasizing these differences, and stressing that morality is the token of humanity, whereby the "universal order" will be restored. Then, equipped with this view, we may be quite certain that animals will be treated in a humane way.

References

Baird Robert M., Rosenbaum Stuart E. (eds.) 1991, *Animal experimentation: The moral issues*, Buffalo: Prometheus Books.

Brophy Brigid 1972, In pursuit of a fantasy, in Godlovitch et al. 1972, 125-45.

Cherry Christopher 1989, Reply to Dolby: The possibility of computers becoming persons, *Social Epistemology* 3, 337-48.

Cottingham John 1978, 'A brute to the brutes?': Descartes' treatment of animals, *Philosophy* 53, 551-559.

Dawkins Richard 1976, *The selfish gene*, New York: Oxford University Press.

Feinberg Joel 1974, On the rights of animals and unborn generations, in William Blackstone (ed.), *Philosophy and environmental crisis*, Athens: University of Georgia Press.

Finkelstein Jacob J. 1981, *The ox that gored*, Philadelphia: The American Philosophical Society.

Frey Raymond G. 1980, *Interests and rights: The case against animals*, Oxford: Clarendon Press.

Godlovitch Stanley, Godlovitch Roslind, Harris John (eds.) 1972, *Animals, men, and morals: An enquiry into the maltreatment of non-humans*, New York: Taplinger.

Godlovitch Roslind 1972, Animals and morals, in Godlovitch et al. 1972, 156-72.

Hettinger Edwin C. 1991, The responsible use of animals in biomedical research, in Baird, Rosenbaum 1991, 115-127.

Harris John 1972, Killing for food, in Godlovitch et al. 1972, 97-110.

Harrison Peter 1991, Animal pain, in Baird, Rosenbaum 1991, 128-139.

Hyde Walter W. 1916, The prosecution and punishment of animals and lifeless things in the Middle Ages and modern times, *University of Pennsylvania Law Review* 64, 696-730.

Kelsen Hans 1945, *General theory of law and state*, New York: Russell & Russell 1961.

Kleining John 1978, Human rights, legal rights and social change, in E. Kamenka and A. Tay (eds.), *Human rights*, London: Edward Arnold, 36-47.

Linzey Andrew 1976, *Animal rights: A Christian assessment of man's treatment of animals*, London: SCM Press.

MacDowell M. Douglas 1978, *The law in classical Athens*, Ithaca: Cornell University Press.

Nelson Leonard 1956, *A system of ethics*, New Haven: Harvard University Press.

Regan Tom 1982, *All that dwell therein: Animal rights and environmental ethics*, Berkeley: University of California Press.

Regan Tom 1991, The case for animal rights, in Baird, Rosenbaum 1991, 77-88.

Singer Peter 1976, *Animal liberation*, London: Jonathan Cape.

Singer Peter 1991, To do or not to do, in Baird, Rosenbaum 1991, 145-150.

Stone Christopher D. 1987, *Earth and other ethics: The case for moral pluralism*, New York: Harper & Row.

Tomkins Peter, Bird Christopher 1973, *The secret life of plants*, New York: Raw & Harper.

Wright Robert 1991, Are animals people too?, in Baird, Rosenbaum 1991, 43-53.

Young Robert 1978, Dispensing with moral rights, *Political Theory* 6, 63-74.

Chapter 3

Moral Dimension of Man
and
Philosophy

Many philosophers throughout the ages have tried to identify a distinctive feature of human being. Probably the most frequently used feature was man's reason or ability to think. However, this ability has been allegedly transferred to the computer and we are left empty-handed, if man is made a machine, or too-full-handed, if machine is made a person. While there are still some criteria uniquely identifying man - such as being a featherless creature with flat nails, or having a specific genetic makeup - they are somewhat unsatisfactory, since they do not seem to convey the essence of human being.

It seems that this characteristic of humanness can be found in the moral dimension of man. When discussing animal rights, Carl Cohen aptly observes that "rights arise, and can be intelligibly defended, only among beings who actually do, or can, make moral claims against each other" (Cohen 1991, 104), that is, beings who participate in "a community of moral agents". It is the moral dimension that becomes a hallmark of beings to whom we can truly ascribe rights, both legal and moral.

In this chapter I would like to present views of some philosophers coming from various traditions and representing many different schools of thought, who agree on one thing: man is a moral being, or, the moral dimension is of primary importance in man and rational dimension is a vehicle through which this dimension develops and acts.

Plato

The idea of the moral distinctiveness of man is by no means new. One of the first philosophical theories which ascribed a prominent position to the ethical dimension of man was given by Plato.

Plato distinguishes three parts of the soul, each of which has a corresponding virtue. Temperance is a virtue of the appetitive part, courage is a virtue of the emotional part, and wisdom of the rational part. Thanks to knowledge, all three parts are held together in a harmonious unity, since wisdom knows "what was the advantage both of each single one and of their commonwealth of the three classes" (*Republic* 442c). Wisdom offers men a glimpse into the eternal world of ideas to acquire knowledge necessary for shaping our lives according to this knowledge which is necessary to reduce, if not escape, the transitory character of the empirical realm. All activities of man depend upon the soul, "upon which neither men nor gods have anything more precious" (*Phaedrus* 241c); the soul, in turn, depends upon wisdom, the source of everything good (*Statesman* 259c, *Timaeus* 46d), therefore, wisdom has the highest value of all virtues, and knowledge is the most valuable pursuit. In this situation, it seems that the rational dimension of man occupies the preeminent position and constitutes the essence of man. However, this dimension is subservient to the ethical dimension, the latter being determined not that much by the structure of man as by the structure and the origin of the world.

In Plato's universe there are two worlds, the transcendental world of ideas, which constitutes true reality, and the world of sensually perceivable objects, which is in constant change. However, everything has only one source, the idea of good. The idea of good is a pre-condition of existence for all things knowable by man. Furthermore, the idea of good is considered to be the source of all ideas, even of the idea of existence. "The ideas are only specific determinations of the good". It is the beginning of all things, "the first principle of all existence ... itself ... underived," and the ultimate goal. "The good is not itself a state of being" (*Republic* 509b). The idea of good is like the Sun not only shining on all things, but allowing them to live, thrive, and grow. From an epistemological standpoint, the idea of good enables cognition, since it "provides their truth to the things known, and gives the power of knowing to the knower ... being the cause of understanding and of truth" (*Republic* 508e). The second goal of knowledge is to know more about the world of ideas and the principle of this world, the idea of good, not simply to satisfy sheer curiosity, but to base our lives so that all institutions of our world are rooted in the reality and truth. Also, knowing the world of ideas makes possible to perfect individual people and society ruled by those people, who have the best knowledge of good, the philosophers. Therefore, perfection of the ideal world can be embodied in this world only when filtered through reason. The better our knowledge is, the better is the chance that our social life is closer to the ideal, and the better the possibility of making the good a part of our life. ⌐The views of Plato on this subject can be summarized in a statement that the ethical foundation of reality causes the rational in man to be in service of the ethical.⌐ To phrase it in a Hegelian spirit, the good as an ontological principle realizes and recognizes itself in the world of appearances through the rational part of the human soul. The ethical principle created the world of ideas and the word of appearances to permeate the latter through man who becomes the nexus between these two words. Thus, the rational in man is subservient to the moral, and man would not be a rational being if he were not founded upon the good.

This principle is exemplified in Plato's idea of happiness. Wise man, a philosopher, can be the happiest person because he is the wisest. But wisdom is not the essence of happiness, but a way to it, since it allows him to avoid extremes and live a full and active life according to the highest moral values. He continues an active life even after death not contenting himself with mere contemplation.

The ontological dimension of Plato's philosophy is substantially reduced in Aristotle's philosophy. Aristotle is much less interested in the transcendental than in the physical. Also, he explicitly expressed the idea that man is a rational being. The rational dimension of man is the highest faculty of man and everything else is subordinate to it, including ethics. Perfect happiness is found in contemplation, especially in philosophy which is a domain that has no other end beyond itself (*Nicomachean Ethics* 1177a17). His ethics is not founded upon the idea of conformity with the good, as in Plato - the world of ideas is rejected altogether. Aristotle mentions the then-popular feeling that a wise man is dear to the gods (1179a24) but this fact apparently has no influence upon moral life of this man. Gods are not much interested in men, therefore men have to strive for the perfection of themselves by themselves. Since no good in general exists, man is concerned about human good residing primarily in the good of reason and in other faculties only insofar as reason influences their order.

From Plato and Aristotle lead two paths concerning the ethics vs. rationality relation, one giving preeminence to ethics, one to rationality.

Augustine

Augustine mentions Cicero and his *Hortensius*, a reading which turned him to thinking about God and immortality (*Confessions* 3.4), but Platonism is in his opinion the best of pagan philosophies (*City of God* 7.4-7) and he embraces this philosophy, and merges it with the Christian religion.

For Augustine man is a complex being. He has soul (*anima*), a vital force, which animals also have. A higher element is rational soul (*animus*) whose highest part is mind (*mens, ratio,* or *spiritus*). He distinguishes also lower and higher reason, the latter (*intellectus,* or *intelligentia*) excelling the former, since intellectual cognition can lead man to God, whereas lower reason is limited to earthly cognition allowing man to create scientific theories.

Augustine is a dualist, and makes a distinction between body and soul which, however, are not accidentally coexisting elements, but elements truly unified. Soul is a superior part of man, and when "soul and body are united, they are called man, and the name is preserved even if they are discussed separately" (*City of God* 13.24.2). The essence of soul must not be confounded with the essence of body, but the nature of man is the union of both, body being its part (*The care to be taken for the dead* 3.5), man is man both in *mente* and in *carne* (*Sermons* 154.10.15) and is defined as "rational substance consisting of soul and body" (*The Trinity* 15.7.11). Place on timeline w/Hume

Like for the ancient philosophers, for Augustine reason is a passive faculty; on the other hand, will is active, therefore, will is the main element of the spiritual life. And because the nature of anything can be expressed only by an active element, nature of man is expressed not by his knowledge but by his will. In this way Augustine parts from intellectualism so characteristic of Antiquity.

God is central in Augustinian philosophy, since whatever exists and whatever happens in the world originates in God. God is an object of metaphysics, epistemology, ethics, and religion. There is no area of human endeavor which should not be God-directed. No independent logic or science exist which do not have God as their goal (cf. also Drozdek 1995).

Man cannot act without God's grace, cannot think and will, and attaining a knowledge of God by the power of human reason, will, and feeling is impossible without a supernatural incentive, without illumination sent to man. It is truth which guides the mind within us and to which we can refer through mind (*The teacher* 11.38-39). This illumination through "inner light" allows man to have a clear understanding (*True religion* 49.96), and it accompanies man even in solving mathematical problems (*The free choice of the will* 2.8.24). This illumination gives intelligence a better grasp on its contingence upon God.

As in Plato, the good is at the beginning, and it is always present, not detached from cosmic, social, or personal events. Therefore, everything that exists, is good. However, for Plato the idea of good appears as an afterthought and occupies only a marginal position in his philosophy. It is an idea of impersonal good which makes possible an explanation of the beginning of the universe. For Augustine absolute good is absolute God; absolute good is not in God - it is God; it is not his attribute, but his essence, identical with his justice and wisdom (*The Trinity* 15.5.7-9), an attribute of a person through whom and for whom everything exists. He is "the good above all goods, the good, whence all goods are derived, the good, without which nothing is good, the good which is good in itself without other goods" (*Expositions on the book of Psalms* 134.6). Since he, as an absolute good, creates the universe, the latter must also be good - "whatsoever exists, is good" (*Confessions* 7.12) - therefore, goodness permeates the universe, the reality of the universe is equated with its goodness, its reality is its ethicality. In particular, matter is also good, since how could God, who is good, create something which is evil?

The problem of evil is solved by not ascribing it a status of substance: evil is a lack of goodness, non-existence of goodness, therefore, evil does not exist, and hence everything can be seen as good. Evil is caused by man when he does not seek God, but the satisfaction of his own desires by attaining perishable things. For good people, to have good life means detachment from the perishable, but evil people find good life in using it (*The free choice of the will* 1.4.10). Thought can join desires in this pursuit, but only free will can direct it there (1.11.21). Therefore, will has to be good will, and "happy is man who wants good will" despising all apparent goods (1.13.28), and "happy people are by necessity good" (1.14.30). Free will is a necessary condition of moral life, it can lead to a good life as well as to an evil life. However, the latter course is unnatural, since free will was created to have man do good, not evil. Therefore, punishment is justified. On the other hand, without free will no deed would be deemed wrong, but it would not be deemed good either. Men would be automata. Hence, "it was necessary that God gave man free will" (2.1.3), and God should not be blamed for giving free will to man who misuses it (2.18.48). Free will makes man a moral agent, but it does not automatically direct him to the good, since it can be "the means of good life or a tool of sin."

Men can act freely only when God influences their intelligence and will. They would be nothing without him, but they do not lose their identity because God's action. God does not destroy their liberty but assures its activity, since will is also a part of the order of causes (*City of God* 5.9-12; 7.30). Spiritual beings cannot be corrupted against their will; the only condition is obedience to God and adherence to his incorruptible goodness. Disobedience leads to sin and suffering. The soul cannot achieve happiness in detachment from God. To desert the supreme good is to act against natural tendencies (*The nature of the good* 7; *City of God* 12.3). Only a soul which is immersed in the supreme good can find fulfillment and happiness (*The free choice of the will* 2.19.52). Man's pursuit of the highest good is not accidental, since the yearning for happiness is inborn to be a guide toward God, and can be fulfilled only in the union with God, the supreme good (*The Trinity* 11.6.10; *The free choice of the will* 3.3.7). Aiming at this good is the goal of the soul, its purpose to live, the seventh and the highest stage the soul can reach. It is not even a stage but "as it were, a dwelling place, to which the preceding stages lead," such as purity of soul and knowledge of the truth (*The greatness of the soul* 4.33.70-76).

Thus, the knowledge of the truth is a precondition of attaining the good, since "only truth allows to know and to have the supreme good" (*The free choice of the will* 2.9.26), therefore, reason is a faculty which is instrumental in attaining it. To this end, reason cooperates with faith. Reason by itself is unable to reach this goal (*The advantage of believing* 1.1.2), reason is not autonomous, it is only one of human faculties which without support of other faculties can achieve very little. Faith precedes reasoning, but is unable to recognize truths without participation of reason, since "it has to be decided whom to believe" (*True religion* 23.45).

Reason is only one of elements of human endowment, which in the long run helps man to reach the highest goal, a "dwelling place," which is a union with the supreme good. All through his life man is oriented toward good, and forgetting this fact leads to chasing chimeras, that is, non-existence, which is evil. Reality is ethical, ethics is hyposthasized, substantiated, whereby moral dimension can be ascribed not only to humans, by to the whole of reality. However, only humans realize that and use the power of their reason to persistently continue in their march toward reality *par excellence*, which is God, which is goodness.

The influence of Augustine on the history of theology and philosophy can hardly be overestimated. His views were undeniably dominant in Western thought up until the thirteenth century, the time of Thomas Aquinas. But even later Augustine's views remained very popular, to mention only Renaissance, philosophers of Port-Royal, and the Cambridge Platonists. Even in the last two centuries the number of works analyzing Augustine's thought is phenomenal.

Thomas Aquinas

Thomas undertakes the difficult task of blending Christian tradition represented by Albert the Great, Jerome, John Damascene, and in particular by Augustine with Aristotle. As a Christian theologian and philosopher he could not accept Aristotle's view on non-existence of the highest good. Aristotle with his intellectualism is his guide in analyzing man limited to the earthly sphere, where humans are first of all rational beings. But in his analyses of the supernatural, Aristotelian elements are considerably limited, and Augustine becomes his teacher.

Man is an intellectual being: he possesses reason which is interested in truth, and will, which is interested in good, and always follows intellect (*Summa theologiae* (1265-74), 1.19.1), thereby being an intellectual appetite (in distinction to sensual appetite). The will acts under the guidance of reason, from which it receives goals to desire and consequently to activate actions leading to fulfillment of its desires. Also conscience is not a special faculty, but an act of reason forming judgment with regard to moral problems, "a certain pronouncement of the mind" (*ST* 1.79.13).

Aristotle says that intellect is the highest power of the soul (*ST* 1.82.3) so that soul is even called by Thomas the intellect or the mind

(*ST* 1.75.2), but this priority should not be understood absolutely. Considered in the abstract, intellect is more excellent than the will. Intellect is self-sufficient, it finds perfection in itself, it already possesses its object.   But the will is characterized by desire, it is outward-oriented, it finds fulfillment in attaining something which is not its own.   Therefore, the will has to strive constantly in order to find satisfaction, and this striving takes place outside of it, whereas intellect's strivings are inside of it.   In respect to material things it is the same, since it is better to know a stone than to desire it, but in respect to divine things the will is superior, "for to will or love God is more excellent than to know him", since the will tends to the perfect goodness, and intellect knows only created goodness (*On truth* (1256-59), 22.11c; *ST* 1.82.3).   Intellect is limited to what is created, and can be outdone by the will if the latter lifts the eyes of its desire above the earth toward perfect goodness, eternity, the uncreated.

This priority of the will over reason should be taken into account when judging goodness of actions springing from will's desire.   Although the wills of different people can be good when desiring opposite things, the reason should direct the will's attention to a broader picture when submitting to the will a particular end to be desired, since "man's will is not right in willing a particular good unless he refers it to the common good as an end" (*ST* 1-1.19.10); and since "the will's object is proposed to it by reason" (*ST* 1-1.19.3), reason should do its best to support will in its maintaining a higher status over reason.   Reason's task is to supply reasonable goals and reasonable solutions based upon a broad knowledge of the situation, of the structure of society and of universe to have the will desire goals which are good, in fact, best possible.   Reason by itself cannot attain these goals, only the will - by leading to action - can.   Reasoning is, therefore, good-oriented, it is directed toward moral good as well as an ultimate goodness, which is God.   Also, reason by itself, although it can give correct advise, is not sufficient for proper behavior.   What is needed are habits formed by the practical application of known rules.   These habits create in man a disposition toward morally acceptable behavior, since, as aptly put, "it is not just knowledge of virtue we seek, but the acquisition of virtue, and virtues are acquired, not by the study of moral philosophy, but by repeated acts of a given kind" (McInerny 1982, 102).

This orientation toward others is not very strongly emphasized by Thomas as the element which extols the will over reason.   Position of reason and rational dimension of man is well within Aristotelian perspective when man is analyzed on the level of his earthly existence. Thomas' effort, however, is directed toward showing that this image of man can be fit in the universe created by God and directed to God, since in him only all creation - including man - can find perfect fulfillment and perfect happiness. In this way, man's rational dimension is subservient to man's moral dimension in the sense that man's goal is to achieve union with the highest good and this good is not a product of reasoning.

First, reason is not self-sufficient in setting goals and directing will. Reason has to function properly and this is possible only when it bases its reasoning process on proper assumptions. Yet these assumptions are not created by reason itself. "Even scientific certainty is not attained without recourse to the first indemonstrable principles" (*Summa contra gentiles* (1261-4), 4.54). Principles of speculative reason, which are "bestowed on us by nature," belong to a special habit, "the understanding of principle," and principles of practical reason belong to *synderesis* (*ST* 1.79.12). Such principles may be "self-evident principles", such as the law of contradiction, or the principle of practical reason saying that "good is to be pursued and done, and evil is to be avoided" (*ST* 1-2.94.2). As self-evident, they are a part of natural law. This formal law has a concrete application in theoretical reason's pursuit for knowledge, since knowledge is a good for reason. With respect to human actions, reason has to include the precepts of the eternal law (*ST* 1-1.21.1).

Man strives to attain the contemplation of truth and the union of soul and body is to facilitate this process (*SCG* 2.83). Contemplation of truth is the end of all human endeavors and everything else is subordinated to this goal: "soundness of body ... freedom from disturbances of the passions - this is achieved through the moral virtues and prudence - and freedom from external disorders" (*SCG* 3.37). However, although this contemplation brings about human happiness, it is not achievable in this life. Contemplation consists of more than the mere knowledge of God, since (almost) all men have some knowledge of God, and of more than cognition of God, attained by faith. But it must be achievable, since we desire such happiness - even more: we are meant for happiness (*ST* 1-2.2.4) - and the existence of a desire which is impossible to reach is unnatural ("nature does nothing in vain," as Aristotle says), therefore happiness can be reached after this life (*SCG* 3.48). This contemplation is a perfect knowledge of God which surpasses natural powers of man. Even increasing these powers is not sufficient, since seeing divine substance can be attained by vision which is not "of the same essential type" as the vision of created intellect. Hence, intellectual power has to be amended by "acquisition of a new disposition ... fittingly called the light of glory," since it endows intellect with sufficient understanding (*SCG* 3.53).

However, it is not all just intellectual operations transferred from divine mind to human mind. Thomas insists that perfect happiness, perfect beatitude can be found in the enjoyment of God. The drive for happiness is an inborn and natural desire, but "the desire for enjoyment of a thing is caused by love of it," therefore man has to love God in order to desire this perfect beatitude. This love is awakened in man by God's love (perfectly expressed in the incarnation) (*SCG* 4.54), therefore, love is the path which spurs in man the desire of the union with perfect goodness, love is the avenue allowing man to become aware of where perfect vision can be found, this perfect vision, "the light of glory," which can make man one with the goodness and perfection. Therefore, although Thomas sees in contemplation of truth the ultimate end for man, so that even "the moral virtues and prudence" are just the means to this end, it is not an expression of pure intellectualism, as it is in Aristotle. Everything hinges here upon the nature of God. Divine essence lies in wisdom, goodness and charity (*ST* 2-2.23.2 ad 1), but it is the goodness and love of God that makes this goal desirable. Would man consider the contemplation of truth his goal if this truth were evil and unpleasant? Would it not be damnation rather than beatitude to seek for such a truth? Truth means love and goodness and that is why man sees in it beatitude and in perfection of his rational power the avenue to this eternal felicity. Man is not just truth-oriented, but good-oriented, this good, the goodness of God, is his goal, this good which is also the truth. Goodness is an ontological principle attaining of which requires goodness as moral principle, the latter being a reflection of the former as its source. Contemplative life is moved by the will using reason to attain the desired goal, knowledge of truth (*ST* 2-2.180.3), therefore it is justified to call contemplation a moral virtue (Gilson 1961, 151). No exercise of science, no theoretical thought or speculative reasoning can satisfy this goal (*ST* 1-2.3.6), only contemplation of the truth.

Man within the confines of the natural world is a rational being.

Reason, as observed already, commands will, it also has conscience under its supervision. Thomas rationalized conscience to the extreme: for him, "conscience is neither sincere intention, nor an affective evaluation of morality of our acts, nor an instinct which surges spontaneously" (Elders 1987, 88). In the natural world man is, so to speak, an Aristotelian being *par excellence*. Human reason reigns here, since it is the highest faculty on which man can rely, and the will, along with sensual appetite, are directed by it. Moral decisions are also intellectualized, and moral virtues used for reaching an end within human limits can be acquired by human actions, although imperfectly (*ST* 1-2.65.2). However, the Aristotelian cast breaks when a theological perspective is taken into account. It is this perspective, which is the most important for Thomas.

Thomas discusses three theological virtues, faith, hope, and charity, which can be infused to man only by God himself. These virtues dispose human mind and will to be directed by God. Through faith, the will seeks God as the most desirable goal. Faith is "supremely necessary for man," since without it man "can neither do nor have anything good" (*Commentary* on *On the Trinity* 3.1); without faith society would collapse, since its preservation requires believing in promises, testimonies, etc., and human reason unaided by faith can lead man astray, since reason is a subject to error as exemplified by philosophers "who searching for the end of human life by way of reason ... fell into ... disgraceful errors" (ib.). Also, "faith gives the speculative intellect a perfection by making it obey the will" (*Debated questions: on truth* (1269-72), 1.7; cf. *ST* 1-2.56.3), whereby faith is a cause of rationality to be subjugated to the moral dimension.

Through hope the will is able to bring down all obstacles leading to God. But the preeminent theological virtue is charity - "the key to understanding Thomas's treatment of the moral life" (Nelson 1992, 70). Man is and should be God-oriented, by being limited to his natural powers he cannot attain the state of beatitude, and "reason ought to be subject to God, and to fix in him the end of its will" (*ST* 1-2.109.8). It is charity, which allows man to attain a state of ultimate perfection, because "it is charity that unites us to God" (*ST* 2-2.184.1). Through theological virtues, God brings about in man seven gifts (such as wisdom and piety) which alter human behavior and make him responsive to God's inspiration (*ST* 1-2.68.1,4). The presence of these gifts results in twelve fruits, such as kindness and moderation (*ST*

1-2.69.4). Also, a person of charity can benefit from the blessings spelled out in the Sermon on the Mount (*ST* 1-2.69).

For Thomas the world is not a combination of static objects; it is a world of acts, events and changes. However, he does not reiterate simple Heraclitean *panta rhei*, since he was constructing a Christian world-view, which is - unlike later in deists - a God-centered system; therefore, reality is not a collection of events in flux, but a system of goal-oriented events, an ordered system, where events develop in a way proper to them. There is, namely, a natural tendency for everything in the world toward better so that both natural and intelligent agents develop for the better. "Every action and movement are for the sake of some perfection" (*SCG* 3.3). Thereby, they tend to the good, since "a thing is good to the extent that it is perfect," which also means that they tend to divine likeness, because "a thing is made like unto God in so far as it is good" (*SCG* 3.24); there is also a natural inclination "to spread ... good amongst others" (*ST* 1.19.2). Hence, an observation that it is "a kind of morality immanent in senseless nature" is to the point (Gilson 1961, 22). Therefore, evil in the world is something unnatural, and, although rampant, evil does not constitute the true essence of this world. Evil, in natural world, is simply an effect of unfavorable circumstances, an accidental diversion from the natural tendencies toward perfection. Evil is not caused by evil, since it "has no cause efficient through itself" (*SCG* 2.41), and "no essence is evil in itself" (*SCG* 3.7). It is good cause which happened to be in a wrong place at wrong time. As for human actions, the resulting evil is caused by faulty reasoning (faulty "judgment of apprehensive power") which submits to the will some good incongruous with the good proper for a given action (*SCG* 1.95; 3.10), evil is a defect in reasoning or in execution of some plan, it is defective action resulting in defective effect.

Man is no exception to this common tendency toward good. Man instinctively wants to achieve perfection, and this perfection can be found on earth through virtuous life directed by natural law and human law, and eternally through subduing himself to the guidance of God spelled out in his law and infused through charity. Man also knows where this perfection can be found and directs his entire rational and affective life toward achieving this goal. He knows that God's "substance is His goodness," "He is good essentially," i.e., by essence (*SCG* 1.38; ST 1.19.1), and only he is good by nature, because only he exists by nature, and is the ultimate goal of all things (*ST* 1.6.3), including man. Therefore, the efforts of the rational dimension are unceasingly directed toward this goal, are subdued to the highest good and, in this respect, to the will, whose object of desire surpasses a naturally possessed object of reason.

In spite of the very strong intellectualist bias of Thomas' philosophy, it can be seen that rational dimension is instrumental in achieving what is really important to the will - the highest goodness. But, among philosophers of the late Middle Ages, Thomas' insistence on the importance of reason was the strongest. Other philosophers and theologians ascribed to reason less importance, in particular, the Franciscan tradition.

For Francis of Assisi, it is only through love that it is possible to attain true knowledge of the object of love, and morality is simply an inner command of love. His subjugation of reason to feeling is called "an immensely important experiment in morality" (Blanshard 1961, 61). For Robert Grossetest moral judgment should be made in accordance to divine law and revelation, and for Roger Bacon, who espoused the same view, an inner moral illumination is possible to at least some men (*Opus maius* (1267), 7.3). For Bonaventure reason is not the highest human power. Man's ultimate beatitude is to be found in an act of love and not in the act of intellectual knowledge. Conscience, through moral illumination, can assess goodness of action. For Duns Scotus the will, free and rational, is the highest faculty of man, not reason, and it gives orders to reason (Bourke 1970, 136, 138-9, 151-2). Will is the essence of the soul. All moral acts are actions of the will (*Ordinatio*, prol., 5.1-2). Knowledge is not the highest good. For William Ockham moral actions can only be "in the power of the will" (*Quodlibet* (1333), 3.13); ultimate end of human endeavors is the highest good, God, whose existence cannot be proven, as in Thomas, but can only be believed, since (centuries before Hume) he questioned the causality principle.

The problem of the essence of man, his characteristics, and his position in the universe reemerges with a new force in the Renaissance. Renaissance is no longer seen as an epoch which 'discovered' man, although it did accentuate certain aspects of man due to new conditions, opening of new perspectives in technology, science, and geography. The Renaissance was not a total reaction against medieval thought, only "the confirmation of some doctrines which were questioned at the end of the Middle Ages. As long as there was no doubt concerning the immortality of the soul, it was difficult to deny the dignity of man and his privileged position in the universe" (Verbecke 1976, 204).   Philosophical reflection of this time focused on man, became anthropocentric, which, in fact, became a hallmark of the Renaissance humanism.   This was a reaction against Aristotelian bent of the late scholasticism which was reflected in the fact that 'divine' Plato was contrasted with this 'beast,' Aristotle, by so many Renaissance humanists (Garin 1965, 11), and even peripatetics admitted that, without Plato, it was impossible to understand Aristotle (Garin 1965, 129).   Renaissance had generally a Platonist twist, very spiritual and religious.   "Renaissance thought was more human and more secular, although not necessarily less religious than medieval thought" (Kristeller 1972, 2), and "the humanists' ideas of human nature in conjunction with their religious ideas ... seem to be a definite characteristic of the Renaissance" (Trinkaus 1970, 763).

Ficino

Marcilio Ficino was one of the central figures of the Quattrocento.   He was a founder and leader of the Academy of Florence.   He translated into Latin Plato's works and the works of leading Platonists.   He zealously propagated Platonist philosophy which he considered a powerful tool in the battle against atheism and in infusing new life to Christian faith.   Not surprisingly, Augustine was also a guide and master (*dux et magister; praeceptor pariter et patronus*, Marcel 1958, 602) for him.   He considered himself "alter Plato" and is called "the philosophical mouthpiece of the Renaissance" (Kristeller 1964, 266). His ethical views are scattered throughout his commentaries on Plato's dialogues, in *Theologia Platonica* (1469-74), in *De Christiana religione* (1474), in many short treatises, and in innumerable letters.

Man is a rational being, but he also has will, "an inclination of the mind toward the good" (Kristeller 1964, 257).   Man's will seeks particular and temporary goods, but is never satisfied, and it can find full satisfaction only in the union with God, an infinite goodness (p. 262), "the good itself" (p. 61).   Similarly, intellect's search for truth and complete satisfaction can only be found in God "in whom alone lies the entire concept of truth and goodness" (p. 262).   A predominant role in this strive for union has love, understood as "the desire for beauty" (p. 263).   Love thus defined, however, has a wider scope and significance, since it eventually transforms soul into the good, that is, into God (p. 264). This is possible, since beauty is radiation of the highest good which is God, and emanates from this good onto the whole of both natural and supernatural worlds. Beauty is like a hook which lifts the soul to the supreme good, an enticement, which leads it in right direction. The orders of love and will are parallel, sometimes indistinguishable: at one point Ficino defines love as "perpetual disposition of will" (p. 270). By doing so, Ficino returns to identification of beauty and goodness, which was held in Antiquity, and was expressed in the composite word *kalokagathos* (ie., *kalos kai agathos*, beautiful and good) not infrequently found on pages of Plato's dialogues (cf. Festugière 1941, 32). Union with God, the goal of soul, can be achieved only through love, not through intellect, whose knowledge of God will always be imperfect. Will causes soul's movement toward its goal, intellect being passive and unable to break out of its limitations. Therefore, love becomes a central category of Ficinian philosophy, especially in later works (Kristeller 1964, 276).

His philosophy of love was actualized to some extent at the Academy of Florence which, for Ficino, was not simply a teaching institution, but "a living community of friends", because he considered a school to be the place for "an intellectual and moral communion between master and pupils" (p. 283, 285). Ficino coined a term for the love between the members of such an intellectual community, Platonic love. This type of love makes our actions other-directed and allows for a moral and spiritual *catharsis*; this, in turn makes proper use of knowledge possible. "As a dirty vessel makes dirty by its contagion whatsoever fluid, even the sweetest, you may pour into it, so a bad mind when receiving knowledge produces malice, not wisdom" (p. 301). Therefore, love is a leading force in all human endeavors, and union with good is the most desirable goal.

Pomponazzi

Although the Platonist trend was very strong in the Renaissance, Aristotelianism was also very popular at many universities, such as the University of Padua, where Pietro Pomponazzi was a professor. Pomponazzi was, in fact, the greatest Aristotelian philosopher of the sixteenth century. *De immortalitate animae* (1516), the most famous of his works, stirred up a very unfavorable reaction at the time. It was publicly burned in Venice in 1517 and in the next year the pope, Leo X, demanded him to recant his views.

On the immortality of the soul is a long answer to the question of Hyeronimus Natalis', whether Pomponazzi believes that soul's immortality can be proven "leaving aside revelation and miracles, and remaining entirely within natural limits" (Preface). The answer is negative. However, in the concluding chapter "revelation and miracles" are used as a definite proof that soul is immortal, and all arguments to the contrary "are false and merely seeming, since the first light and the first truth show the opposite" (ch. 15). In spite of his repudiation of these "false and merely seeming" arguments, Pomponazzi devotes fourteen chapters of his book to them. What is interesting for us are the arguments that he uses against the eight objections commonly cited by advocates of the immortality of the soul.

The first of these objections sees danger in the view of mortality of the soul because it closes the way for man to reach happiness, that is, contemplation of the Highest Being (chs. 13, 14). In his answer, Pomponazzi indicates that men share in three types of intellect: theoretical, practical, and productive. There are some theoretical principles known to every man, such as "something exists or does not exist," and there are practical principles which are commonly known, principles concerning customs, public and familial affairs, since "to each it is given to know good and evil". As for productive intellect, it is clear that everyone would perish without it, and even animals possess it. Perfect theoretical intellect is attainable by but a few. Practical intellect is "truly fitting for man" and it can become perfect in everyone. Therefore, all men should be good, but all men do not have to be philosophers or scientists. The moral dimension of man is an indispensable condition of preservation of humanity, and the world would disappear if everyone theorized and speculated. Since society is an organism which requires harmonious interaction of all its members, everyone participates by bringing in some share of theoretical and also productive intellect. However, practical intellect has to operate in everyone. Therefore, "all ought to be righteous and good. But to be a philosopher, a mathematician, or an architect is a particular end."

The moral dimension constitutes a foundation of individual and social life, other dimensions have a secondary importance. Belief in the mortality of the soul does not have to destroy the moral dimension, however, a pessimistic and somewhat unnerving tinge of this view can fade away if it is firmly founded on the conviction of the priority of the moral dimension in man. It can be even maintained that mortality preserves virtue better than immortality, "for hope of reward and the fear of punishment appear to savor of certain servility which is contrary to the nature of virtue". Virtue by itself is the best reward and it does not require any substantiation in the form of immortality of souls. Vice, by itself, is the worst punishment. Truly, "Pomponazzi's ethical doctrine is the most radical part of his philosophy" (Pine 1986, 107).

The high priority given to morality in such diverse philosophers of the Renaissance as Ficino and Pomponazzi are expressions of assigning priority to will over intellect, so characteristic in this epoch. In the disputes between voluntarism and intellectualism, which led to the disputes between Thomists and Skotists in the fourteenth century, and in the Renaissance, the voluntarist position clearly prevailed. This position was directed against Aristotelian contemplationism and stressed the activism of man, as well as openness to others, and the social character of man based upon his morality. Therefore, Pico della Mirandolla "was passionately interested in problems concerning love and the will. The problem of knowledge seems to have interested him less" (Dulles 1941, 130). Coluccio Salutati, and then Bernardino de Siena say that "the will is the king of our mind" (Garin 1965, 37). Contemplation is of little value, says Paolo Paruta, because we can reach God through contacts with people, and "all that is necessary is to make allowance for the good deeds done by one man to another in the course of civic life" (Garin 1965, 182); we become citizens of heaven not speculating about nature but acting morally, says Aleksandro Piccolomini (Garin 1965, 174).

Hobbes

Much of the discussion concerning human nature in the seventeenth and eighteenth centuries was spurred by Thomas Hobbes, in particular by his *Leviathan* (1651). He attempted an explanation of all things in terms of matter in motion, in purely deterministic fashion.

There is no greatest good, but "felicity is a continuall progresse of the desire, from one object to another," and "a generall inclination of all mankind" is "a perpetual and restlesse desire of Power after power" (ch. 11). Good is simply an object of man's appetite, evil an object of his hate (ch. 6). If now, in the presence of laws and regulations man behaves so contemptibly, how much worse it must have been in the natural state when no laws existed. As constructed by Hobbes, man in his hypothetical natural state was selfish to the extreme. Men were born equal, meaning with equally selfish desires and appetites, which could lead only to perpetual conflicts and skirmishes. Then reason discovered a fundamental law of nature that it is forbidden for man to destroy his life, which implies "that every man ought to endeavour Peace" (ch. 14). Therefore, the only solution was to renounce some desires in order to preserve the rest by creating state with laws regulating behavior of its citizens. "This is the Generation of that great Leviathan, or rather ... of that Mortall God, to which wee owe under the Immortall God, our peace and defence" (ch. 17). This peace and defence, this security and well-being of all, leads, in Hobbes' view, to development and progress which would have been impossible before the establishment of laws and a sovereign and absolute state.

The fundamental law just mentioned is one of nineteen "Lawes of Nature dictating Peace" and "the Science of them is the true and onely Moral Philosophy" (ch. 15). Although man is not by nature good and cannot find rules of peaceable conduct in himself, he can, nevertheless, find them in nature. Therefore, Hobbes' nature is saturated with moral laws waiting for discovery and because everything in his world is strictly determined, moral progress is a cosmic law, and nature allows man to make this progress by unfolding him laws dictating peace.

Hobbes' man is a determined automaton, driven by noxious feelings leading to self-defeat if not regulated by the state. It was a rational being, but his reason was at the service of his hyper-egoism. In this sense Hobbes' man was primarily a moral being, but his morality was amoral, pitch black and destructive. Man is a moral being but his morality is of a regrettable quality. Nature, however, hides in its womb the laws that allow man to reach a higher plane of development by disclosing these laws to reason. Nature, therefore, possesses a content which - when poured into man's moral ability, through his reason - leads him out of condition, in which, as Plautus said, *homo homini lupus*.

The Cambridge Platonists

One of the first reactions against the Hobbesian view of man and society came from a group of philosophers now known as the Cambridge

Platonists. The philosophy of this school continued the Platonic tradition which was very firmly rooted in European philosophy, due in particular, to Augustine. Their main concern was ethics, however, they were very firmly opposed to the Hobbesian relativism of moral principles and continued the Platonic tradition of realism and absolutism in ethics.

Ethics is not based on fluctuating principles depending on the will of a sovereign, or Leviathan, but on rules that are eternal and immutable: "Moral Good and Evil, Just and Unjust, Honest and Dishonest ... cannot possibly be arbitrary things, made by will without nature", says Ralph Cudworth in *A treatise concerning eternal and immutable morality* (published posthumously in 1731, 1.2.1). They are so basic and immutable that even God cannot change them by the power of his will (1.3.1). Goodness is superior to wisdom, and goodness constitutes the essence of God, who can be viewed as "an infinite Circle, whose inmost Center is Simple Goodness, the Rays and expanded Plat thereof, all Comprehending and Immutable Wisdom, the Exterior Periphery ... Omnipotent Will" (1.3.8). Just as Plato, for whom Good was an idea of ideas, Cudworth sees goodness as an element which determines the activity of man's reason and will. Since all things are created by God, the cosmos is, ontologically, based upon goodness. Moral principles cannot be acquired by perception; they are innate, since otherwise "there could be no such Thing as Morality" (4.6.4).

Henry More is more specific about these principles which he calls *noemata* or moral axioms, and in *Enchiridion ethicum* (1667) he lists twenty three of them. All these *noemata* are "immediately and irresistibly true, [and] need no proof" (1.4). More also introduces a faculty distinct from reason and will, called the boniform faculty of the soul, which allows man to distinguish and to relish in the highest good (1.2.5). However, More is not consistent in his discussions of this faculty. It is identified sometimes with ethical sense (Lichtenstein 1962, 67) or with moral sense (Sorley 1921, 121). Through this faculty man is not just automatically following the moral principles which are imprinted on his soul, but he also takes pleasure in responding to this highest call. Goodness is a foundation of the universe and only man is able to consciously follow its rules thanks to the reason with which he is endowed. Through intellectual love, which is contained in the boniform faculty (2.9.16; cf. Austin 1935, 41), man can also attain the highest happiness by following the twenty three moral axioms. Intellectual abilities allow man to recognize the central principle of the universe, the principle of goodness, and to become one with goodness by following its ways which are posted with road-signs within the self.

Cambridge Platonists emphasized the role of reason. Reason is a source of true knowledge, which "cannot possibly spring from Sense" and results "from an active Power of the Mind" (Cudworth, *Treatise* 4.6.12). "Reason in a Good man sits in the Throne and governs all the Powers of the Soul in a sweet harmony and agreement with itself" (John Smith, *The excellency and nobleness of true religion* (1745), ch. 2), and "to go against Reason is to go against God" (Benjamin Whichcote, *Moral and religious aphorisms* (1703), 76). However, this is not an unqualified rationalism, if the ontological structure of the world is also taken into account. As indicated, this ontological structure rests upon goodness, and, therefore, is directed toward this goodness. Cambridge men have moral and religious dimensions always on their minds, and the rationality of man serves only the purposes of these dimensions. Human reason is the means of knowing the truth about the world, that it is a creation of the good Supreme Being and thereby saturated with his goodness.

Reason is not perfect (it rarely is), and it cannot remain an undisputable guide. It has to cooperate with the heart and be strengthened by rightful conduct: "nothing is the true improvement of our rational faculties but the exercise of the several virtues of sobriety, modesty, gentleness ..." (Whichcote, *Aphorisms* 541). Reason cannot remain detached from this activity and from human will by restraining itself to the theoretical sphere. Reason was not brought into being for that purpose, and if it turns its back on practical activity, it distorts both the *noemata* imprinted on it and the clarity of its perception. "Men want not so much means of knowing what they ought to do, as wills to do what they know. Practical knowledge differs from all other knowledge, and excels it," says John Smith in *A discourse concerning the true way or method of attaining divine knowledge*, sec. 2 (cf. also Cassirer 1953, 41, 124-6). Knowledge of good presupposes a goodness of man, which can be developed and exercised by applying the moral precepts inscribed in the soul and by looking at the "candle of the Lord" with which we are all endowed. Therefore, man can be good if goodness is known clearly and distinctly, to use the Cartesian phrase, which, in fact, John More also uses after listing all his *noemata*.

Reason is edified by conduct, the latter being good if reason is properly functioning ("right reason"). Man is a moral being all throughout his life, making reason a means of infilling this life with morally acceptable content.

Locke

Locke's philosophy exercised a big influence at the end of the seventeenth and in the eighteenth centuries. Locke did not write any treatise on ethics, but his entire philosophy was motivated by ethical and religious issues, since he wanted to establish a firm basis for the Christian religion and the ethics which springs from this religion. Knowledge was a central issue in Locke's writings, in particular, in his *Essay concerning human understanding* (1690). But Locke also analyzed epistemological problems, attempting to show that the human mind is able to prove all the basic tenets of Christian religion and ethics by starting with a clean slate, with no preconceived knowledge, just with *tabula rasa*. He was convinced of the veracity of Christian religion and, at the same time, of the power of reason. Since Christianity was true and reason was the supreme power able to lead man to truth, then they could not contradict each other, if the power of reason was properly exercised.

Reason is a distinctive faculty of man, it distinguishes him from animals, and only reason can be trusted as a reliable guide for all man's needs in life, including rules of conduct and ethical precepts. Moral principles cannot be detected in mind, they have to be proven with mathematical precision (*Essay* 1.2.4). Man is clearly a rational being and his ethics is only an outcome of his rational abilities.

However, Locke's rationalist outlook is not consistent in all respects. Not infrequently he concedes to the priority of revelation in ethical matters. For example, he asks rhetorically, "whether one coming from heaven in the power of God ... be not likelier to enlighten the bulk of mankind, and set them right in their duties, and bring them to do them, than by reasoning with them from general notions and principles of human reason?" (*The reasonableness of Christianity* (1695), in *Works* vii, 146), and states that "the Gospel contains so perfect a body of ethics that reason may be excused from that inquiry, since she may find man's duty clearer and easier in revelation than in herself" (*Letter to Molyneux*, March 30, 1696, *Works* ix, 377).

He criticized innate ideas but not innate faculties, primarily natural reason which is inborn (*Essay* 2.11.10). Morality is concerned with complex ideas, called modes, which are "made very arbitrarily, made without patterns, or reference to any real existence" (3.5.3); they are the mind's creations. The main problem with this solution is that "morality built on mixed modes could not give rise to moral obligation at all" (Lamprecht 1962, 78). They cannot even lead to any action. Reason can submit some practical solutions, but by itself, it is power-less. Conscience cannot serve this purpose either, since in Locke's system it is a purely rational faculty: it is "our opinion or judgment of the moral rectitude or pravity of our own actions" (*Essay* 1.2.8). Therefore, Locke considers a desire as "the chief, if not only spur to human industry and action" (2.20.6). In that regard, reason can be referenced by passions and emotions, and can "suspend the execution and satisfaction of ... desires" (2.21.48), but reason's voice can be overruled (cf. Petzäll 1937, 52-5). As he admitted in *Thoughts concerning education* (1693), reason is unable to tame some pains without resorting to some help, and "our first actions" are "guided more by self-love that reason or reflection" (secs. 107, 110). Man is not an embodiment of rationality, and if he is restrained only to the rational dimension, he becomes a purely abstract construct having no counterpart in reality. Even if reason is ascribed the highest position among human virtues, it does not constitute a leading force in human life, and the throne on which it is seated can be encaged by faculties which philosophers view as less admirable.

Locke's ethical rationalism (as any rationalism) did not succeed to its fullest extent and tracing Locke's efforts to explain man in a rationalistic fashion only shows a relaxing of his opinions, starting with the unrealistically strict work *Of the conduct of the understanding* (published posthumously in 1706) and ending with the following editions of *Essays* and *Thoughts concerning education*. Locke's man is a rational being, but the goal of this rationality is to bring forth moral principles as clearly as possible, based upon Christian religion which is presumably retrieved by reason from reality. Man is a ra-tional being in order to become moral being. This morality is so im-portant and so prevalent, in fact, that it breaks the bounds of rationality by referring to revelation, if necessary. To be rational is to engender morality, but this result is known in advance, rationality is teleologi-cally bound to reach this goal. Therefore, man only truly becomes man if the moral principles are adequately formed and pursued and if his rationality is crowned with a moral dimension. Otherwise, the whole rational enterprise is wasted cause.

However, Locke's epistemological empiricism had shown a way of solving the problem of the moral dimension in man, the way indicated by Thomas Burnet. In *Third remarks...* (1699) he observes that there may be a faculty in man which is able to perceive the goodness or badness of an act as senses perceive tastes, shapes or colors of objects. This innate and natural faculty, called natural conscience, can distinguish "one thing from another in Moral Cases, without ratiocination". It may be difficult to explain of what the goodness of an act exactly consists, but it is equally difficult to explain in words the difference between different colors, smells or tastes (quoted in Tuveson 1960, 48). This faculty, since it is analogous to senses, can properly be called moral sense, and this is the name that was used by Shaftesbury and later popularized by Hutcheson. Therefore, Burnet can be seen as an originator of the concept of moral sense, but not of its name (Tuveson 1960, 51, but cf. Grean 1967, 277).

Burnet stressed more than others the analogy between sensory perception and moral perception, however, traces of this analogy can be found in other authors of the seventeenth century. The statement of John Hartcliffe in his *Treatise of moral and intellectual virtues* (1691) can serve as an illustration: "it is a secret impression on the minds of men, whereby they are naturally directed to approve some things as good and avoid others as evil ... the former appear beautiful and lovely, the latter ugly and deformed" (quoted in Voitle 1955, 26). This secret impression, or moral sense is able to direct man to see his actions in the proper light, thanks to an unspecified mechanism, which Hartcliffe would agree, is similar to sensory perception. In any case, the concept of this faculty was more widely discussed in the seventeenth century (and also in the next century), and therefore, seeing Burnet as the originator of moral sense may not be entirely justified.

Shaftesbury

The eighteenth century continued its interest in philosophical anthropology and the thinkers of this century attempted to solve the problem of the essence of man by analyzing him primarily as a moral being or as a political and economic agent. These two approaches were not necessarily separate, since the moral and social dimensions of man were sometimes intermingled and supportive of each other. The moral trend of this century opened with Anthony, Earl of Shaftesbury, mainly by his *Characteristics of men, manners, opinions, times* (1711), which had eleven editions in this century and was read in all of Europe.

Shaftesbury was against the Cartesian identification of man as a thinking subject and against the mechanistic concept of man. Desiring also to protect an individual man from supremacy of state and church, he envisioned man as an autonomous and independent being who can find in himself the foundation for action. To be sure, he was not the only philosopher who had this goal. Leibniz, for instance, has tried it in his monadology. Shaftesbury wanted to show the links which bind man to society and religion based upon the unmovable foundation of his human specificity. In the spirit of the Cambridge Platonists and as a reaction against Hobbes he wanted to restore the dignity of man, man's natural goodness and morality which is *in potentia* in each and every man and is developed through his life.

Man by nature is able to make moral decisions, and in this respect, "'tis best to stick to common sense and go no farther. Men's first thoughts in this matter are generally better than their second." Too much reflection can even be harmful, since upon the first thought man would oppose a certain morally tainted action, whereas "speculative men find great modifications in each case: many ways of evasion" (*Characteristics* 2.4.1).

Sidgwick calls Shaftesbury "the first moralist who distinctly takes psychological experience as the basis of ethics" (Sidgwick 1949, 190), although, in this respect, he is indebted to the Cambridge Platonists. For Shaftesbury, "it is by affection merely that a creature is esteemed good or ill," therefore, first of all, affections should be classified accordingly (4.1.2.1). He distinguishes three types of affections: natural affections, "which lead to the good of the public," self affections, which are self-oriented, and neither of these, or unnatural affections (4.2.1.3). This list also reflects the weight that Shaftesbury assigned to these affection: natural, or social affections are first on the list and they are undoubtedly the most important. Shaftesbury scoffs at the entire philosophical tradition which attempts to build ethics on selfishness alone. Those attempting it are "narrow-minded philosophers" who "have a mighty fancy to reduce all its [heart's] motions, balances, and weights, to that one principle and foundation of a cool and deliberate selfishness." Shaftesbury mentions explicitly Hobbes, lord Rochester and La Rochefoucauld as such philosophers (2.3.3). These social affections, termed by other writers benevolence, are basic to man and constitutive to his nature, otherwise, how can even the existence of a basic social units, such as a family, be explained?

Compared to other philosophers who attempted to rebut Hobbes, Shaftesbury's emphasis on the affective nature of morality was unique. Hobbes based his analyses on two premises: man was evil by nature, and moral behavior springs from affections. Many critiques of Hobbes (the Cambridge Platonists, Reid, Price) denied both premises, whereas Shaftesbury used the second premise to fight against the first. "The subsequent history of benevolism offers ample proof that his response was the more effective one" (Voitle 1955, 27).

Morality characterizes each man, and as it is impossible to find a totally good person, so it is impossible to find someone absolutely immoral. Even "a ruffian who out of a sense of fidelity and honor of any kind refuses to discover his associates ... has certainly some principle of virtue" (4.1.2.4).

Unlike Locke, Shaftesbury says that the moral principles of justice and honesty are innate. In this, man's nature reflects nature itself, since virtue "is really something in itself, and in the nature of things; not arbitrary or fictitious" (5.2.3). Principles of right and wrong are unalterable and also accessible to human reason and moral sense. As it is remarked, for Shaftesbury "one must *be* good before one can really *know* good" (Grean 1967, 220). Being good is a matter of affections and also the heart, which must "rightly or soundly affect what is just and right", whereas knowing is undertaken by the mind which "readily discerns the good and ill towards the species or public" (4.1.2.3). An attitude toward good and wrong is expressed by the whole of man, by his rational and emotive faculties, since man's morality, expressed in particular in his social affections, is a distinctive feature of humanity. This moral dimension is summarized by Shaftesbury in the term "moral sense" (4.1.3.2), which was used very frequently by later moralists, in particular, by Hutcheson. In the spirit of Thomas Burnet he explains that, as through the senses, the "natural sense of a sublime and beautiful" can detect proportion or beauty in shapes or colors, so an apparent difference, according to the regularity" can also be perceived in "behavior and actions" as it presents itself to the moral sense (1.1.2.3).

Rationality is essential in the process of becoming a moral being, but reason's primary task is to help man develop the social affections and moral virtues necessary for the promotion of social well-being to the fullest extent possible. Without reason, all moral virtues are dormant and become activated only after man's "reflecting faculty" is added to the being's "wanting reason." Only then will this being be "capable of virtue, and [will] have a sense of right and wrong" (4.1.3.3). However, pure rationality, devoid of the moral dimension, is not desirable, since if man loses the "natural sense of the odiousness of crime and injustice," that it, conscience, then he becomes "miserable in the worst way" (4.2.2.1). Both dimensions have to be present and operating in order to attain the fullness of a perfect moral sense. Thus, albeit Shaftesbury does not forget the "reflecting faculty" in his discussion of man's morality, it is true that "almost the whole stress is laid on the benevolent affections and the 'Moral Sense,' while but little is said of ... the share which the Reason takes in estimating the character of our acts" (Fowler 1883, 71). It is also true, as already observed by Butler (Grean 1967, 243), that the sense of obligation springing in particular from conscience, did not receive sufficient attention in his discussions. However, it is clear that for Shaftesbury, man is primarily an ethical and aesthetical being making the best use possible of his reason to develop and edify these dimensions. This tendency can be observed through all the alterations and amplifications Shaftesbury introduced in his works. Moral faculty becomes more credible and authoritative, and it is more and more associated with the sense of beauty, whereby its credibility is strengthened (Voitle 1955, 37).

Butler

One of the most prominent figures in the moral philosophy of eighteenth century England was Joseph Butler, whose ethical views are outlined in frequently reprinted *Fifteen sermons preached at the Rolls Chapel* (1726) and *Dissertation upon virtue* (1736). Butler distinguishes four separate parts of human nature, all hierarchically ordered: 1. "particular affections, passions and appetites" (Preface to *Sermons*), 2. "benevolence, or the love of another" (*Sermon* i), 3. self-love, and 4. conscience, also called moral reason or moral sense. Conscience has an absolute supremacy over other faculties and its decisions concerning a course of actions are binding. Self-love and conscience do not conflict "and always lead us the same way" (*Sermon* iii). Their analysis indicates that "we were made for society, and to promote the happiness of it" (*Sermon* i). Recognition of good by conscience is possible, since, generally, there is nothing inherently bad in man, and each element of his nature was designed for a good cause. We may not recognize it, because of our imperfection, and "however perfect things are, they must even necessarily appear to us less perfect than they are" (*Sermon* xv).

Butler leaves no doubt that man is primarily a moral agent, and conscience is his highest faculty. The importance of conscience is also seen in Butler's apparent disinterest in analyses of the cognitive faculties of man. He only mentions this faculty in passing, however, even a casual analysis of the intellectual dimension of man is conspicuously missing. Butler also seems to assign little importance to the inner workings of conscience, mentioning only that it "can be considered as a sentiment of understanding, or a perception of the heart; or, which seems to be true, as including both" (*Dissertations*). He was primarily interested in indicating the position morality occupies in man and showing that our moral dimension is what makes "brute creatures" human.

Hutcheson

The problems analyzed by Shaftesbury were further developed by his students. One of them was Francis Hutcheson, whose *An inquiry into the origin of our ideas of beauty and virtue* (1725) became very popular in his time. Following Shaftesbury's ideas, he tied ethics with aesthetics, beauty with good. Hutcheson's analyses of beauty had strong bearing on his vision of ethics. In his *System of moral philosophy* (1755) he opposed the views which derived ethics from nature, society, and egoistic predispositions of man. He wrote about the moral sense (or moral faculty) which determines human actions.

Moral sense is self sufficient and no corroboration of its decisions is required on the part of reason or education. Moral sense is a prior and fundamental faculty of man, whereby ethics cannot be reduced to any other area. Reason is submitted to moral decisions, since it is "too slow, too full of doubt and hesitation to serve us in every exigency" (*Inquiry*, quoted after Scott 1966, 224) and only supplies the means to realize the goals determined by morality. Moral judgments are made by moral sense, motivated primarily by benevolence. Moral sense allows man to be benevolent toward others, treating them, to use Kantian terms, as ends and not as means. Like Shaftesbury, Hutcheson was sure that even the people we would judge immoral are not devoid of moral sense, since, for example, "Robbers ... have their sublime moral Ideas of their Party, as Generous, Courageous, Trusty, nay Honest too" (*An inquiry concerning moral good and evil*, 1725, sec. 3.5); they are guided in their actions by benevolence and good intentions misguided by false reasons and misjudgment of historical situation.

Hume

Hume was more aware than others about the fundamental significance of philosophical analysis of man for all areas of science. He starts his *Treatise of human nature* (1739) with the statement, "it is obvious that all sciences are in some way connected to the human nature" - even mathematics. Therefore, the proper exercise of science necessitates an adequate knowledge of human nature. And Hume's interest was concentrated upon man as a moral being.

Hume devotes the second book of his *Treatise* to an extensive analysis of passions. Passions are "reflective impressions" resulting from original (i.e., perceptual) impressions, or from ideas. There are direct passions which "arise immediately from good or evil," such as fear, joy, or hope. Indirect passions have also some qualities absent from direct passions. Some examples of indirect passions are love, hatred, and pity (2.1.1). Passion is a primary existence being no copy of anything else and containing no representative aspect; passions, volitions, and actions are "original facts and realities, compleat in themselves" (3.1.1). Reason is concerned with "the world of ideas," while will places us in reality. Therefore, demonstrations of reason and acts of will are far apart (2.3.3), and there is no conflict between reason and passion: reason reflects reality, it is knowledge about reality, and passions are reality, therefore, passions cannot conflict with reason. Reason is confined to finding connections between ideas - according to the principles of resemblance, contiguity, and cause or effect (*An enquiry concerning human understanding* (1748), sec. iii) - and therefore, it can only direct judgments and have no influence on human actions. Actions "may be laudable or blamable, but they cannot be reasonable or unreasonable" (2.3.3).

Hume also opposes the traditional view of the subservience of will to reason. "Reason alone can never be a motive of any action of the will," and "it can never oppose passion in the direction of the will" (2.3.3). "Reason is, and ought only to be, the slave of the passions, and can never pretend to any other office than to serve and obey them" (2.3.3). This statement, called by Norman Smith (1941, 45) Hume's "fundamental maxim", can be found later in Poincaré's statement that science uses the indicative mood and not the imperative mood. This is one of the most controversial elements of Hume's philosophy and many attempts are made to take the edge off this statement (eg., Glathe 1950, 9-13; Mackie 1980, ch. 4). The fear is that reason as a slave of passion will be arbitrarily tossed by figments of our feelings and thus reason will turn into unreason, and rational thinking into irrationalism. We should not forget, however, that there is a division of labor in Hume's philosophy: reason's area is truth, but morality is driven by passion. Passions have no impact upon decisions of reason in the sense that they do not convert truth into falsity and vice versa. "Reason is the discovery of truth and falsehood" (3.1.1), it *is* an embodiment of this truth-searching faculty, and morality does not deprive it of this capacity. Hume's intention can be found in the closing statements of his *Treatise*, when he says that "the most abstract speculations concerning human nature ... become subservient to *practical morality*, and may render this latter science more correct in its precepts and more persuasive in its exhortations" (3.3.6). All our intellectual endeavors are subdued to the interest and well-being of man, all their values should be assessed by moral sense. Science has no priority over morality, it is this practical morality which directs the course of human intellectual activity and any other type of activity. Also, it is the morality of man which distinguishes humans from other creatures, and not reason. Reason does not distinguish man from animals, since the latter are characterized by "reasoning that is not in itself different, nor founded on different principles, from that which appears in human nature" (1.3.16). Neither language is this distinctive mark, despite a very strong Cartesian tradition in this matter in the eighteenth century. It is this moral dimension which distinguishes man from animals on both individual and social levels.

All this led to Hume's vision of ethics, which is active and able to influence our actions thereby going "beyond the calm and indolent judgments of the understanding ... Morals excite passions, and produce or prevent actions. Reason of itself is utterly impotent in this particular. The rules of morality, therefore, are not conclusions of our reason" (3.1.1). Moral distinctions come from moral sense which by feeling, not by judgment, allows us to have impressions or sentiments indicating what is good or evil. These impressions are experiences of pleasure connected with virtue and experiences of unease in presence of vice (3.1.2). Moral sense, therefore, determines our attitude towards a particular act, since all facts concerning this act can be equally known to different people, and yet they will assess this very same act differently. "When Nero killed Agripina, all the relations between himself and the person, and all the circumstances of the fact, were previously known to him; but the motive of revenge, or fear, or interest, prevailed in his savage heart over the sentiment of duty of humanity." We know the same facts as Nero, and yet we are revolted by his act "for the rectitude of our disposition" (*An enquiry concerning the principles of morals* (1751), Appendix I).

But the actions are not guided only by the moral sense that can be found in all people. Man is a social being and society has its say as well, in particular, by demanding justice. Sense of justice and injustice does not appear from nature, but is a result of impressions which "are not natural to the mind of man, but arise from artifice and human conventions" (3.2.2). Justice is an artificial virtue which allows for the continuation of social life. There are also natural virtues - such as courage, compassion, gratitude, friendship, fidelity, unselfishness, etc. Although artificial, justice is a virtue, and "the sense of its morality is natural" (3.3.6), since justice maintains harmony in society and it is certainly in our best interests and it also produces satisfaction and pleasure.

There are, therefore, two moral orders in Hume's system which was an attempt to overcome the extremities of other positions: not all virtues are natural, due to the nature of man's constitution, as Shaftesbury would maintain, and not all virtues are merely conventions created arbitrarily to lubricate the gears of society, as Hobbes wanted it. Man is on the borderline between nature and society; as part of nature he is generously endowed with a moral dimension which, in fact, elevates him above nature. But nature, so to speak, was interested in man as an individual and did not supply man with sufficient equipment to allow him to conduct a social life. Therefore, society has to fill this gap by creating a set of artificial virtues, in particular the virtue of justice. The same division can be found later in Bergson's distinction between open and closed societies.

Rousseau

"If we consider human society with a calm and disinterested eye, it seems at first sight to show us nothing but the violence of the powerful and the oppression of the weak" says Rousseau in the Preface to his *Discourse on the origin of inequality* (1754). This was an appalling fact to Rousseau and he wanted to discover whether it was an unavoidable situation, a necessity arising from the nature of man. Society was based on inequality, therefore, the question is, is man really a social animal as Aristotle stated and so many thinkers of the Enlightenment repeated? Despite the emphasis of this age on scientific, philosophical, and technical progress, and despite the admiration of the philosophers for the achievements of human reason, Rousseau looked primarily at the society which brought this progress rather than at the progress itself. And as we can read it in all his writings and see it in his life, his opinions and statements concerning the inequality which pervaded his society were anything but calm and disinterested.

His plan was to go to the beginnings of humanity, even if these beginnings have to be constructed and based on "hypothetical and conditional reasoning" (Rousseau 1971, p. 177), in order to discover the driving forces of man and the essence of his humanity. Was man then the same as he is today, and if not, what happened to bring him to today's deplorable situation?

In the natural state, man lived as animal, free, happy and thoughtless, knowing no language, waging no war, free of diseases, having plenty of food. He lived in a continuous present, taking no heed of the future, forgetting yesterday, not interested in technical progress and even when making a discovery, doing nothing to preserve it for others. Social ties were non-existent and no fixed residency was established. But man was not just an animal, he was a free being, and while not being entirely driven by instincts, he did have the liberty to resist the voice of nature or acquiesce to it. Another distinctive characteristic of man was his faculty of improvement, of productivity, which brings birth to all other faculties. Man was neither bad or good, since no "moral relations" existed between men, nor any duties (p.

199). "[S]avages are not bad, precisely because they don't know what is to be good; for it is neither the development of understanding, not the curb of the law, but the calmness of their passions and their ignorance of vice that hinder them from doing ill" (p. 201). From this we can infer that man has predispositions towards being good because of the calmness of his passions. This despite the very pessimistic outlook of Hobbes who saw only perpetual anarchy and *bellum omnium contra omnes* of the natural man. This natural calmness is, in fact, derived from "the only natural virtue" man has, the virtue of pity, which allows for the feeling of sympathy and identification with those who suffer. This feeling was perfect in the natural state and it was at least severely unnerved, if not completely stifled, under the influence of reason. "It is reason that engenders self-love, and reflection that strengthens it; it is reason that makes man shrink into himself" (p. 203). This is this sentiment of pity which allows species to survive and not "finespun arguments" of philosophers.

This stage of wild and mild brutes ended with the growth of the population. A very primitive society emerged based upon familial bonds, and, to some extent, on tribal ties.   Families begin to live together under one roof, bringing forth "the sweetest sentiments" of "conjugal and parental love" (p. 216). When the original wildness is shaken off, people assemble for leisure and amusement. Man is now, more than before, gentle and meek, distant from "the stupidity of brutes, and the pernicious enlightenment of the civilized man" (p. 217). It was "the happiest and the most durable epoch" (p. 218). But when men started to undertake tasks which surpassed the abilities of one person, then this harmonious state was broken, and this can be traced to the beginning of agriculture and metallurgy. In modern terminology, the neolithic agricultural revolution marked the end of the natural state, and the invention of metallurgy in the Bronze Age made this end irreversibly fatal. Private property was created, the natural state ended, and civil society slowly emerged. "The first man, who after enclosing a piece of ground, took it into his head to say, *this is mine*, and found people simple enough to believe him, was the real founder of civil society. How many crimes, how many wars, how many murders, how many misfortunes and horrors would that man have saved the human species" if he restrained from foolish actions (pp. 211-212).

This was a paradise lost and not to be found again, at least not in the same form. Mankind is "no longer able to retrace its steps" (p.

226), "nature does not retrograde and it is never possible to return to
the times of innocence and equality once they are behind us"
(*Rousseau, judge of Jean Jacques* (1772-6), dialogue 3). History
cannot be reversed at will, and Rousseau did not even propose it.
Therefore, Voltaire's deriding remark that Rousseau's *Discourse on
inequality* was a call to return to walking on all fours was simply ill-
conceived.

After losing the natural state, man's pity was severely suppressed,
his reasoning faculties, however, become superior, as man became a
rational being. Happiness and equality are not, however, a lost cause.
It is possible to regain them by transforming society into a different
form in which the most appalling elements of contemporary society are
continually being uprooted and replaced by new social relations. This
continual spiral of social improvement would bring humanity to
freedom. Reason led to the loss of freedom, reason will be used now
to regain it. This can be done, says Rousseau, in a society outlined in
The social contract (1762) and his two novels (*New Héloïse* (1761) and
Émile (1762)).

The new society allows man to become fully human by his regaining
freedom, in particular "moral freedom, which alone renders man truly
master of himself". He also acquires advantages impossible in the
natural state: "his faculties are exercised and developed; his ideas are
expanded; his feelings are ennobled; his whole soul is exalted" (*The
social contract*, 1.8). Since society cannot be abolished, freedom can
be found in the total dissolution of all individuals in society, because
"each giving himself to all, gives himself to nobody," an outcome being
"a moral collective body" (1.6). This higher level of social development makes man a consciously moral being. In the natural state, "prior
to reason," he was endowed with a natural morality, the feeling of pity.
By becoming a member of society, an individual acquires a sense of
duty toward others. Reason elevates man to a fully human level,
freedom is regained, and humanity develops. Man was by nature a
moral being, if only unconsciously, but now, using reason, he becomes
a social moral being. It is all possible, since society, despite its long
grip on man, did not destroy his natural innocence and goodness (cf.
Starobinski 1971, 33). The ages of inequality and injustice were not
completely unnecessary. Man returns to himself only when atoms are
transformed into a truly harmonious composite, "on the level of self-
consciousness, the wholeness of preconscious man" is reestablished

(Bernstein 1980, 66).

It is interesting to observe a reversed parallel with Hobbes, who, like Rousseau, constructed a hypothetical state of nature and an ideal state. However, for Hobbes, man was originally immoral, selfish and destructive, for Rousseau he was peaceful, basically good, and a happy man. For Hobbes, the state introduces order by also introducing laws, for Rousseau state destroys everything, making man wiser but worse. However, almost dialectically, this undesirable state can be changed and new state formed where goodness and morality are restored.

Kant

Kant was a heir of Hume who interrupted his "dogmatic slumber" (preface to *Prolegomena* (1783)) and of Rousseau who restored in him a "belief in the common man." In a marginal note Kant wrote: "I am myself by inclination a seeker after truth. I feel a consuming thirst for knowledge and a restless passion to advance in it, as well as a satisfaction in every forward step. There was a time when I thought that this alone could constitute the honor of mankind, and I despised the common man who knows nothing, Rousseau set me right. This blind prejudice vanished; I learned to respect human nature, and I should consider myself far more useless than the ordinary working man if I did not believe that this view could give worth to all others to establish the rights of man". And, like Rousseau, he gave a preeminent position in human thinking and acting to ethics, holding, after Rousseau, the view that what is essential in man "consists in man's ethical and not in physical nature" (Cassirer 1961, 21), but, unlike his predecessor, Kant did it in painstainkingly systematic fashion presenting a magnificent system whose heart is ethics.

Kant wanted to defend man's dignity by showing that he is not one of the cogs of a deterministic universe ruled by mechanical laws. In a world depicted as a big machine, there is no room for freedom, goal determined action, and morality. The laws of this world determine all actions and no moral consideration can influence it. In order to defeat this world view, Kant, had to explore first the nature and limitations of scientific cognition, thereby showing that science is unable to encompass everything within the confines of its categories. Science has only a limited scope, and scientists are just "artificers in the field of reason" (*Critique of pure reason* (1781), A839/B867). Kant believed that for "the necessary practical employment of reason" it is essential "to take away from this reason "its pretensions to transcendent insight" (Bxxx). The highest goal of human reason should be the study of the "whole vocation of man" - the study of morality (A840/B868). Thus, the study of theoretical reason precedes the study of practical reason. But it should be stressed that there are no two reasons indwelling a rational being, but only two manifestations of one reason: "In the final analysis there can be but one and the same reason which must be differentiated only in application" (*Foundation of the metaphysics of morals* (1785), preface). Therefore, it may be more proper to say that there are two types of reasoning, theoretical and practical, one concerned with knowledge, the other concerned with action.

Theoretical and practical reasons may, however, have conflicting interests which have to be resolved, since consistency "is a general condition of having reason" (*Critique of practical reason* (1788), 1.2.3). These conflicts are resolved under the guidance of practical reason; that is, practical reason (ethics) has a priority over pure reason (cognition). However, he starts with the traditional assumption that highest status of reason alone can recognize truth and goodness. But he did not, like many predecessors, assume that theoretical thinking has a supreme value that makes man more human; but moral practice, or a good will does. Kant even "argued that of itself theoretical activity is neither unconditionally nor intrinsically good; it is valuable only insofar as it enhances moral practice and offers morally permissible maxims of happiness!" (Sullivan 1989, 97). Therefore, practical reason is the highest arbiter in the case of any conflict and its decisions can nullify even the decisions of theoretical reason. The only end having an absolute value is good will, i.e., good moral character. Objective goodness of things presupposes moral goodness and "everything good that is not based on a morally good disposition ... is nothing but pretense and glittering misery" (quoted in Sullivan 1989, 108). Thus, a pure science not guided by moral sense is unacceptable. Science, for its own sake, can do more harm than good, and the search for truth is justified only by results, that is, by the morally acceptable results it leads to.

Not only science, but all actions have to be guided by moral principles (maxims) of a universalist character and absolute value. Practical reason provides the formal specifications of such a principle that is then filled by rational agent with a concrete content. This

Moral Dimension of Man

formal principle is the famous categorical imperative and one of its formulations states that we should "act so that you treat humanity ... always as an end and never as a means only" (*Foundation*, ch. 2). Thus, this imperative is a formal principle whose content depends on conscience, and subsequently on the moral norms of a given time and place. Kant's categorical imperative specifies the condition of certain norms and not the norms themselves. It is a meta-rule, which filters all the rules submitted by people's consciences.

It is worth noting that this principle applies not just to humans, but to all rational beings, as indicated by Kant's remark that "man and, in general, every rational being exists as an end in himself not merely as a means to be arbitrarily used by this or that will" (*Foundation*, ch. 2) This remark is consistent with the approach Kant uses in his philosophy, namely with considering everything on a level as abstract and as detached from empirical facts as possible to establish principles of an absolute validity and not contingent upon the empirical state of affairs. Therefore, "the metaphysics of morals is really pure moral philosophy, with no underlying basis of anthropology or other empirical conditions" (A842/B870). Only after being fully established, can this pure moral philosophy be applied to man: "it borrows nothing from the knowledge of him (anthropology) but gives him, as a rational being, *a priori* laws" (*Foundation*, preface).

Rational agency is always goal oriented, hence a rational action is invariably intentional, that is, a realization of certain goals. Rational activity is impossible without morality, causing a rational being to be also a moral being (even if his morality is immoral). Rationality and morality are two aspects of the same being. Therefore, a person is a rational and free being, intentionally conducting his actions under moral law. As such, a person has an objective worth and an unconditional dignity. In this sense, animals and plants are not persons and are not ends in themselves. We, as people, have moral obligations with regard to them, but these are really obligations to ourselves. Hence, animals should not be unnecessarily harmed or subjected to cruel experiments, since it is harmful to our moral disposition. Also, Kant would certainly find the supposition that a machine might be considered a person simply distasteful.

Ethical problems were also an important part of Kant's historiosophy and social philosophy. Man and his well-being was a central idea in all of Kant's analyses. Well-being was understood as the realization of freedom on earth, allowing man to peacefully and harmoniously develop his potentials for common good. These analyses also show Kant's idea of man that was not altogether rosy, as a lofty categorical imperative might suggest, and that Kant did not lack a sense of realism with regard to human nature. However, since he was able to go beyond the phenomenal layer of human reality, he could also picture man in a dignified way and was full of optimism about the future.

In his *Idea for a universal history with cosmopolitan intent* (1784), Kant shows that man is striving through wars and calamities toward a state of freedom which would be fully embodied in a just civil society where mankind would be able to develop all man's faculties. All these calamities are, however, indispensable, since without them no progress would be possible: "thanks be to nature for his quarrelsomeness, his enviously competitive vanity, and for his insatiable desire to possess or to rule, for without them all the excellent natural faculties of mankind would forever remain undeveloped" (Kant 1977, 121). In this upward progress, ethical principles will become reigning forces among men, since right now "we are highly civilized by art and science ... but much [is needed] before we can consider ourselves truly ethicized," and "the idea of morality" can be realized only after a long process of rectification of people and states (p. 126). Man ultimately wants to create a state of truly free spirits, i.e., of humans who in their lives realize the ethical ideal contained in the categorical imperative. Kant, man of Enlightenment, saw the state as a means for the realization of moral ideals, but for him, it was rather *malum necessarium*. Only in Romantic conceptions did the state become a fulcrum for history, perhaps best exemplified in Hegel, for whom the state is a realization of freedom and, therefore, the history of freedom is basically the history of the state (in particular in his *Philosophy of history*). To Kant, history is the history of man, and a progression toward freedom, where freedom, not happiness, is the goal (*Presumable origin of human history* (1786)).

In *Religion within the limits of reason alone* (1793) Kant says that man has three predispositions: to self-preservation, to the development of humanity and to the development of personality. However, man also has evil propensities, possibly innate, possibly acquired. Kant's statement that man is evil means that man is conscious of the moral law but occasionally chooses to deviate from it, perhaps driven by passions and inclinations which contradict the moral law. Moral predisposition is an innate element and therefore even the most wicked man does not repudiate this law. Man is a frail creature, not entirely good, but neither is he entirely evil, and for man, "who despite a corrupted heart yet possesses a good will, there remains hope of a return to the good from which he has strayed" (Kant 1977, 392). Moral perfection, or the state of freedom, freedom from evil, is not only possible but attainable and, furthermore, "it is our common duty as men to elevate ourselves to this ideal" (p. 396). It is a common effort, no individual man can achieve it. This effort can lead to establishment of an ethical commonwealth, a realm of virtue. This unifying effort is also a guarantee that the evil can be extirpated from man, since "the species of rational beings is objectively, through the idea of reason, destined for a social goal, namely, the promotion of the highest good as a social good" (p. 407). It was an optimism that Kant never renounced despite not so optimistic developments on the political scene in his times. Kant's deeply rooted conviction about man being predestined to good allowed him to be faithful to this optimistic vision of history.

Fichte

In spite of the fact that he ascribed priority to practical reason, Kant devoted his main work to theoretical reason and his Copernican revolution is associated with the problem of cognition. Moreover, the problem of morality, although important for Kant, was only one of the tenets of his philosophy. For Johann Gottlieb Fichte, on the other hand, the problems associated with practical reason were a primary concern. Fichte analyzed all other problems in this light, since, as he stated, "practical reason is the root of all reason" (*The vocation of man* (1800), 3.1) Because the problem of ethics permeates all of his works, it is very appropriate for his philosophy to be called a system of ethical idealism (Copeland 1965, 52).

This prevalence of the moral dimension in his philosophy can be seen at the very outset, when Fichte determines the goal of philosophy and of principle upon which philosophy should be based. This discussion may be found in his two introductions to *Wissenschaftslehre* (1797). For Fichte, one type of philosophy is arbitrarily based only on the "inclination and interest" of the philosopher, that is, on "what type of man one is" (*First Introduction* (1797), sec. 5). Therefore, the problem of choosing between idealism and dogmatism, the latter retaining Kantian notion of thing in itself, depends on moral maturity: for a morally mature philosopher idealism remains the only avenue, since it stresses the subject, the self, the ego. Therefore, due to his practical approach to philosophy, the self becomes a foundation of Fichtean philosophy. For Fichte, the world is an arena of the self's activity; the ego, including its moral attitude toward the world, is the principle of being, not what is outside thereof. This activist view of philosophy goes as far as proclaiming the priority of the act over substance. Substance is the product of an act, and an act is primary form of being. This activity is a metaphysical foundation for the world, not substance, and philosophy should determine the first acts, not the first facts. In this way, man does not depend on the natural world, he is reduced to the moral consciousness. The outside world has no meaning inherent to it, no meaning independent of the self. The only reason the world exists is to become a part of the process of the self's striving for perfection, it exists only for the sake of man's moral call. Man cannot become perfect without overcoming obstacles, and reality gives him an opportunity to do this. The world exists to allow an individual to become perfect.

To become free is a moral command. All acts are good only because they are acts; what is evil is an inaction, the lack of an act. An act can only be done by a free agent, therefore, if an act is determined by circumstances it is not an act any more, it is a passive happening. What is not of the self, is not an act. Hence, the root of evil is the loss of freedom and being led astray by circumstances. Therefore, a slave is guilty of his state, since he does not act at his own accord. Slavery, to be sure, was criticized by Fichte as morally evil, however, he saw it as the result of a particular moral attitude rather than as a set of historical circumstances. Fichte felt that awakening a moral protest in people was the primary means of solving this social problem.

All men want to reach an ideal state in which they may be fully responsible for their actions. This is man's moral calling. Fichte was against eudaimonism in ethics, which sees personal happiness as the most desirable goal, and he was also against utilitarianism, which treats moral values instrumentally. In *Some lectures on the destiny of the scholar* (1794) we read that the concept of happiness is derived from the moral nature of man and it is not true that man's destiny is the moral good attained through happiness. "What brings happiness is not good, but only what is good makes one happy. Without morality no happiness is possible." Man's ultimate goal is to subdue everything which is not rational and to rule over it according to his laws. Although for humans this goal is unattainable, man should constantly bring himself closer to it, whereby his goal becomes perfecting himself indefinitely. "He exists in order to be morally better and better, and to make everything around him perceptually better, by which he makes himself gradually a happier being."

Fichte's moral stance can be found also in his historiosophy. The history of man is a gradual realization of the idea of freedom. Fichte was one of the few German philosophers who claimed that French revolution was morally right, although he did not accept its terror and its violent methods. He viewed the latter as a result of neglecting a good moral education. In *Characteristics of the present age* (1806) he assumes that mankind is endowed with rationality, which means that mankind is autonomous, but it also has a calling. The existence of mankind is associated with the realization of a moral ideal. Rationality is what is valuable for all men. Throughout the history of mankind, rational principles manifested themselves in a variety of ways. Accordingly, Fichte distinguishes five epochs. In the first epoch, reason was simply identical with instinct because its activity was not free, and did not recognize its causes. In the next epoch, men's actions were directed through the authority of a ruler or an ethical system. In the third epoch (Fichte's time), individual's wishes and desires, regardless of their truth were the driving forces in man's actions. In the next epoch, men acknowledge the weight of rationality and make it a guide for the actions. Finally, in the last epoch men are to act morally and rationally not only on account of external orders, but freely and spontaneously (*Characteristics*, lecture 1).

Fichte made a distinction between morality and law before Kant. Although distinct, these areas are not unrelated, the legal principles were morally obligatory for man. Law was designed to guarantee an individual freedom to exercise control over their own body and property. Law did this through coercion, and paradoxically, men were coerced to be free, an idea also found in Rousseau. This, however, seems to contradict Fichte's statement in *Some lectures on the destiny of the scholar* that man should not force any other rational being to become virtuous, wise, or happy. But Fichte is a realist and does not exclude the possibility of a tension between freedom and coercion. He admits that we should make use of someone else's influence upon us when striving for perfection, since interaction is the path to social perfection which is our calling as social beings, and unavoidably involves some element of coercion. In each man there is a social instinct and living in isolation causes that man not to be fully human. We all are to strive for bringing into existence a perfect society, using governmental apparatus while knowing that state is a temporary means which will eventually wither away (this thought can later be found in Marx and Engels). In this ideal state "pure reason will be commonly held the highest judge" (but this certainly does not rule out the possibility of mistakes and even mischievous behavior). But before this ideal state arrives, the state exists and individuals have to fulfill certain duties towards it, as Fichte explains in detail in *Foundation of the natural law* (1796). In the state, each human is at the same time a man and a citizen. In what does not pertain to the state, man is totally free, and the state is the means of ascertaining this freedom. In his freedom, man attains morality. However, passing through the stage of the state is a historical necessity.

Fichte did not confine his ideas to theoretical discussion, he wanted to bring into being some of his ideas, like Plato's philosopher. In Jena he took a stand against three secret student fraternities (*Orden*) which, he said, limited the moral autonomy of an individual (Kuderowicz 1963, 20-1). He ascribed considerable weight to the problem of education in which he saw a means of solving social, political and religious conflicts. This can be seen in *Addresses to the German nation* (1807-8). Man for Fichte is good by nature and possesses moral laws which are etched in his conscience. "A common view of man as an egoist by nature ... is completely false, since something does not emerge from nothing, and further development of basic instinct cannot transform this instinct into its opposite; hence, how could education implant morality in child, if it were not in this child before education?" (*Address* 10). Education has to take for granted a moral instinct in the child, in order to find its expressions, and to gradually develop it using proper incentives. "The principal rule says that this instinct has to be directed to the only object proper to it - to moral domain." Education was aimed at kindling these good sparks hidden in man, and at building an attitude of moral responsibility and moral consequences for one's actions. An educator's aim is to bring a pupil to the state of loving the world. "Intellectual education was always, for Fichte, a means to moral development, never an end in itself" (Castle 1962, 202). In his

educational ideas, Fichte draws on a Swiss educator, Johann H. Pestalozzi. However, unlike Pestalozzi, he gives the state the leading role
in the educational process.

Moral education was an important goal in Fichte's practical
application of his system. The entire system is saturated with respect
for, and consideration of, the moral dimension of man; the rational
dimension of man was seen as an important means of actuating the true
vocation of man, which is rendering himself to the service of men.
Since - as stated in *The vocation of man* - striving for the good is more
important than striving for truth: if will directs itself toward the good,
then understanding can grasp the truth. If the former is neglected and
only truth is searched for, then only "skillfulness is developed aimed
at vain and void refinements" (ch. 3). Reasoning was powerless in face
of correctly formulated views. The assessment should be done by
conscience, since only conscience was able to decide for sure, on which
side is moral good. Conscience was an unmediated perception of the
highest value. Each individual, as endowed with conscience, was able
to make such a decision. This decision should always be "obeyed in
silence" and never criticized (*The vocation of man*, 3.3), since - as
Fichte assures us in his *System of ethics* (1798) - conscience never
makes and never can make any mistakes. It is not surprising that
Fichte held moral education in such a high esteem. The development
of the moral dimension of man is certainly a desirable and lofty goal
and it is important that this development constituted the core of the
system of the man, whom Johann Wolfgang Goethe called "one of the
most illustrious men in the world" (Fischer 1900, 189).

Scheler

Max Scheler occupies a prominent position in twentieth century philosophy. He concentrated his efforts on developing phenomenological
ethics. Edmund Husserl, a founder of phenomenology, was interested
mainly in very theoretical issues, such as phenomenological methodology; but Scheler had a more practical interest. He developed such
themes as: an analysis of ethics, the ontological foundations of ethics,
the cognition of values, and the social aspects of ethics. The main
work of Scheler, "the most brilliant German thinker of his day"
(Bocheński 1966, 140), was *Formalism in ethics and non-formal ethics
of values* (1913-16), which "unquestionably represents the chief

contribution to ethics in this century" (Frings 1965, 103).

Kant was an intellectualist: although practical reason has priority over theoretical reason, all processes still take place within the realm of reason. In particular, morality is confined to reason alone, and the emotive dimension of man has nothing to do with moral decisions. For Scheler this was unacceptable. Reason has a very important role to play in the life of man, for moral considerations, however, it is of little value. Scheler, unlike Kant, discusses the world of values into which man can have insight using a special intuition. Knowledge of the moral reality is a primordial comprehension of the world of values which exists independently of both man's thinking and volition and the empirical experience of good. Therefore, Kant's ethics is formal, built upon a purely formal law of morality, a categorical imperative, Scheler's ethics is non-formal (*materiale Wertethik*), founded upon the world of values and an immediate cognition or intuition of this world. Scheler's point of departure was a world of values independent of empirical reality, not a world of immutable rules.

Values are independent of empirical circumstances, they are not an abstraction of empirical concepts, otherwise they would lose their independence. Also, values cannot be extracted from a moral system accepted in a certain historical situation by a particular society. Society is changeable, but the world of values is immutable. The world of values is primary to any of its historical manifestations. Values cannot be found in man's self, since this would give them the mere status of subjective validity. Values are objective entities, ideal objects which constitute a self-standing world, and through them things can become worthwhile. A value is, to use Scheler's analogy, like a color which exists independently of the object having this color. Color is not definable in terms of all of the objects which have it (as it would be defined in a distributive mode in the context of set theory). Color can exist regardless of whether or not there are any objects of this color.

The world of values is hierarchical, and values fall into four classes: pleasure values, vital values, spiritual values (the aesthetic values, the distinction between right and wrong, and the value of knowledge for knowledge's sake), and the values of the holy. This immutable hierarchy is based on the equally immutable and eternal forces of God's love. This order of values is reflected in the hearts of all men, whereby the heart is not a "chaos of blind feelings ... [but] an articulated counterpart of cosmos of all possible worths of love, ... a microcosmos of the world of values" (Scheler 1957, 361). Therefore, although the ontological status of values is different than that of human reality, man may have knowledge of this world. But values cannot be accessed through the cognitive processes used in the empirical world. A new kind of cognition has to come into play.

Man's knowledge of values comes through a cognitive channel of a non-rational nature. The dualism of reason and sensibility is "the ancient prejudice" and it should be dismissed as erroneous and ground- less. The phenomenology of values is independent of logic. Scheler makes in this context a reference to the philosophers who ascribed cognitive powers to non-rational faculties: Augustine, and Pascal with his *raisons du coeur* and *logique du coeur*. Scheler writes about inten- tional feeling (Scheler 1973, 257). There are different levels of emotional feelings, and Scheler makes a distinction between the lower type of feelings, such as sensual, or vital, which are called functions, and the higher type of feelings, called acts. The highest level of emotional life are acts of love and hatred, through which moral values become a reality for man.

Love is the basic power of man, the foundation of any action. In the primacy of love, the theoretical and the practical find their unity. The apriorism of love constitutes an irreducible foundation for other apriorisms, including those of cognition and will (Scheler 1973, 64). Love has a creative role in the value comprehension, since love discloses the values in this process not simply following feeling and preferring them (Scheler 1973, 261). Theoretical knowledge does not precede the process of loving. The ultimate foundation of love is in the domain of values. In this domain, God is the highest value, and human love is participation in an infinite love of God, which enables man to gain knowledge of the world of values. "Love is even the essence of God himself, who is the highest good and as such the source of all values, and therefore love, as the participation in God, can comprehend all values in their objective essence" (Przywara 1923, 42).

Man's essence lies in his self-transcendence, in his movement beyond himself. Love is the force that causes this self-transcending movement, since love is a true emotion, that is, a motion, which directs man toward higher values and an ideal of man. Love allows man to gain knowledge, empowers man to see the world of values and enables man to know an inexhaustible world of truths; "*love* is always an *awakener* unto knowledge and willing - even a mother of the spirit and the reason itself" (Scheler 1957, 356). The extent and depth of man's knowledge and his willingness to act is determined by his love, by his *ordo amoris*, since love directs the self to the outside, and knowledge cannot be gained without breaking the monad of the self. "Man is an *ens amans* before he is an *ens cogitans* or an *ens volens*" (Scheler 1957, 356). *Ordo amoris* is a foundation for all moral acts, the basic moral formula (*sittliche Grundformula*) which determines the moral dimension of man (Scheler 1957, 348). Understanding *ordo amoris* allows one to see "the simple basic outline of his heart (*Gemüt*), which deserves more to be called the center of man as a spiritual being than cognition and will" (Scheler 1957, 348).

Love is a prerequisite for each of the three types of knowledge that Scheler distinguishes - that is, control knowledge, knowledge of essence, and salutary knowledge. The higher is the form of knowledge, the stronger is the participation of love in the cognitive process. For example, the knowledge of essence requires humility and love of an absolute value. Salutary knowledge, which is a bridge to the highest good and the foundation of being, is only possible by love, since this highest good, God, can be known solely through love, and also, "of all kinds of cognition, knowledge of God is most intimately linked to moral progress" (Scheler 1960, 265).

A person is a man acting morally, a man responsible for his acts. The concept of responsibility is rooted in the experience of the person, it is not created inductively from outward acts and external analysis. This concept is rooted in an immediate knowledge of an autonomous act, its moral value and not in a mental connection between a fulfilled act and the self. A person can be assessed according to his ideal value-essence (*Wertwesen*), i.e., his value directed intentions. Although the concept of person includes his ability to think, the moral dimension constitutes its core and essence. Without this dimension, man is not a person, although he is still a man (Scheler 1973, 488).

The moral nature of man is important, since men are not isolated individuals, but elements of society, of a "community of persons" who are bound together by their common values and feelings of solidarity, which resembles Fichte's concept of culture. "Under the reign of the solidarity principle everyone feels and knows society as a whole in himself and perceives his blood as a part of blood circulating in it, and his values as components of values found in the spirit of community" (Scheler 1955, 140). This feeling of solidarity is clearly based on the universal system of values which are accessible to phenomenological intuition. The social ties founded on emotions and feelings were more important to him than the ties implied by social, political or religious institutions and organizations.

Scheler makes a similar observation with regard to history: "The soul of history is not the actual event, but rather the history of the ideals, value systems, norms, and forms of ethos with which men measure themselves and their practical activities. ... Accordingly, the history of model persons, their origin and transformation, is the real core of this soul of history" (Scheler 1957, 268). Since model persons are windows through which the rays of the world of abstract values penetrate our world, then, in the long run, these values are the real engines of history, which bring about all development in history and society. The presence of such models is essential to the moral development of other people, since people usually follow examples rather than norms and rules (Scheler 1973, 583). Norms are powerless if people who give them are not living and loving exemplifications of goodness.

Although values are absolute, and different truths are eventually all derived from realm of ideas, they are not perceived in the same way by all societies in all times. Each historical situation creates its own window through which it has an inkling into the truth. However, this window may not be perfectly transparent, and the view may not only be partial but also distorted. Scheler talks about anticipatory schemata, through which men see reality and which differ in different societies and historical situations (1960, 198). It is a "sociologically conditioned perspective of interests which selects the concepts, meanings and eidetic images" and which skews man's perception of reality (p. 429). These schemata are derived from a certain form of essence-perception and from a certain form of ethos.

Ethos is a particular view of, or rather feeling for, the world of values. Man's ethos can consider some values only, which can result in narrow ethical views. Values are actualized and concretized. They are lived and made alive through ethos in its various historical forms. Different historical periods are characterized by different ethos which may give a good insight into the values or may obfuscate and distort them. This falsification is brought about by different causes and Scheler discusses resentment as one of them, in *Resentment and moral value judgment* (1912, see Scheler 1955).

Scheler is exceptional is his treatment of values which are

independent of the empirical and the conscious. For positivists, values were preceded by acts of valuation, for pragmatists, they were relative and adaptable to situations. Scheler attacked utilitarianism as an erroneous ethics which "subdued the goal to the means" (Scheler 1955, 129). Utility should be derived from the good and be subjugated to the good. Utilitarianism is a short-sighted philosophy which focuses upon the here and now, upon immediate gain and immediate use, and which will eventually lead to the alienation of man. Alienation, however, should be eradicated and Scheler saw an acknowledged hierarchy of values as the way to escape this alienation. Humans are able to perceive values and consequently realize them - realize them through love. The moral dimension of man should always be brought to the fore, since it is only through ethics that knowledge is possible and true historical progress is realizable. Suppressing this dimension only leads man on a downward path.

Fromm

A theory which is directed against the ethical dimension of man is Freud's version of psychoanalysis. For Freud man was basically a part of nature, driven by biological drives or instincts and with a social layer imposed on him in the form of the *superego*. The *superego*, an internalized external authority, was composed of expectations from parents and the culture as a whole, which, if violated, led to punishment or even rejection. Therefore, out of self-interest, rules were obeyed. And further, rules were always unwillingly obeyed, since moral rules, along with higher virtues, were not grounded in what is truly human, thus they were hypocritical. This limited view of morality was overcome by Erich Fromm, a psychologist and philosopher, whose vision of ethics can be found in *Man for himself* (1947). He distinguishes two types of ethics and conscience: authoritarian and humanistic.

People ruled by authoritarian conscience do not feel guilty, but afraid. Freud's *superego* belongs under this category, however, *superego* "is only one form of conscience or, possibly, a preliminary stage in the development of conscience" (p. 149). Humanistic conscience, on the other hand is to be "the voice of our true selves which summons us back to ourselves, to live productively, to develop fully and harmoniously - that is, to become what we potentially are.

It is the guardian of our integrity ... the voice of our loving care for ourselves" (p. 163). This resembles Kant's statement from his *Idea for a universal history with cosmopolitan intent* that "the supreme objective of nature" is "the development of all the faculties of man by his own effort" (Kant 1977, 121). To the nature of man belongs the principle that "the power to act creates the need to use this power and that the failure to use it results in dysfunction and unhappiness" (p. 220-1), since "freedom and spontaneity are the objective goals to be attained by every human being" (p. 223).

Fromm sees man as more than a creature driven by libidinal forces and endows him with more of a human dimension. Biological forces are not the basic forces of life anymore, but "decency, love, and courage" (p. 227). Man is not drowned in an ocean of drives which make him one of the animals, while humanness consists of being more prone to become neurotic not only because man turns his back on these drives, but because a human being cannot be reduced to the biological. Man "is the only creature endowed with reason ... man is the only creature endowed with conscience" (p. 234). Reason and conscience obey the forces that life rests upon - decency, love, and courage - with conscience calling man back to himself. Conscience is concerned with the existence of this goal, the reason is concerned with the strategy of attaining it, and intelligence is concerned with tactics (p. 108). Conscience, however, has a broader vision, it sees further and wider. Reason is the means of realizing what the voice of conscience whispers.

"Man is not only a rational and social animal," but also a "producing animal." "Not only can he produce, but must produce in order to live" (p. 91). In this respect, reason is a guiding force, since without reason this production (*sensu largo*, not just material production) would be very short-lived. However, this human aspect of man, man's ability to produce, and the need of doing it, can be discovered by conscience, by man's ethical dimension. Only conscience can bring an awareness that the meaning of life can be brought about by man himself "by unfolding of his powers, by living productively" (p. 53). Man has to discover his humanity first in order to realize it, otherwise his production could not be distinguished from the productivity of an automaton which only produces because it was so programmed.

However, there is one tendency in *Man for himself* which is signaled already in the title of this treatise on ethics: it is an emphasis put on the self only, to the extent that others appear to be just a means of achieving the egoistic goal of the self, which is burgeoning the self's creative powers. Humanistic conscience, Fromm says, is "the expression of our true selves", "man's self-interest and integrity, while authoritarian conscience is concerned with man's obedience, self-sacrifice, duty, or his 'social-adjustment'" (p. 163). It seems, then, that the sense of duty and self-sacrifice, let alone social adjustment, are undesirable elements of conscience, and that the self should entirely rely upon the desire to express its true essence. But what happens if this true self is completely asocial, aggressive beyond measure, stingy and possessive, or sadistic? Should such a self be allowed to develop itself in the name of not suppressing humanistic conscience? There is no other solution, since the only alternative is the authoritarian ethics which has no place in an ideal future world (p. 170-1). It is, however, rather disquieting that in the same breath Fromm lists Protestantism, Fascism and the parent-child relationships in our culture as examples of authoritarian ethics (p. 156). It seems some assumption has to be made about human nature in order to prevent the collapse of a system where self-interest rules.

Man is endowed with two powers which allow him to break beyond himself: the power of reason and the power of love, with care, responsibility, respect, and knowledge characterizing the latter power (p. 104). Love is the love of humans, hence the self, being a human is also an object of love. Love of myself is not the same as selfishness (contra Calvin and Kant), since I can love others only if I am able to love myself. Also, love for others is not a self-depravation (contra Freud). As Butler before him (in *Sermon* xi), with respect to self-love and benevolence, Fromm shows that love of myself does not contradict with love of others, but strengthens it. "Love of one person [for example, of myself] implies love of man as such." (p. 134). Fromm has much more to say about this power of love in his *The art of loving* (1956). There, he is more explicit about the fact that "love is an active power in man; a power which breaks through the walls which separate man from his fellow men, which unites him with others" (p. 20). Expressing one's productivity means expressing the power of love by being responsible and respectful. It is the act of opening oneself to others and growing oneself by giving and caring for others. The more caring, respectful, and responsible we are, the more we grow, the stronger our conscience is and the better fulfilled our sense of life is. The moral dimension is in Fromm's approach blended with love, since for him the creative expression of love for others becomes the highest value of humanistic conscience, this conscience that stirs man to best exteriorization of his productive powers, the power of love, and the power of reason.

Conclusions

This chapter discussed views of philosophers as divers as Platonists and Aristotelians, who were interested primarily in ontology, or in social issues, but one theme unites all of them: their view of the primacy of the moral dimension. This moral dimension was understood differently by different philosophers, however, one element remains the same in all of them: the moral dimension pertains to the problem of good, and the rational dimension functions as a tool which allows man to acquire this good, be united with it, or live for its purpose. All philosophers agree that there is a rational dimension in man - reason, intellect, rational reasoning - but it is just a vehicle to serve our moral purposes. This subservience of reason to the moral dimension of man can be expressed either as subservience to other human faculties whose function is to direct man in moral behavior or direct him to the good as such: it can be the will, conscience, or moral sense. This subservience can be expressed in the view that all activities of reason should eventually lead man to union with, or to understanding of, the good (Platonic tradition, Thomas). In this view, man is a moral being who uses all his faculties to reach a good, either of earthly or supernatural magnitude. Sometimes it is not even either-or, since both kinds of good can be included. Thomas is an example, since, for him, morally good living is a stepping stone which leads to the attainment of the supernatural good, God. Man stands out above other beings by having reason, but possession of reason is not a goal in itself. If it is, man becomes self-enclosed, self-oriented, as it is the fact in Aristotle's views, or driven by forces of non-moral nature, which are beyond his control, such as social forces (Marxists), serving self-interests (utilitarianism, pragmatism), or even instincts (Freud). However, when seen as a tool of our moral dimension, reason acquires proper proportion.

It is very important and interesting that philosophers from such different backgrounds saw the need to place reason in a position which shows that *homo* is really *sapiens*, that is, man's *sapientia* does not crush other dimensions in order to elevate reason to unreasonably high position, but it does make it an apparatus in the service of goals determined by morality.

The philosophers presented in this chapter are only examples of thinkers who see, in one way or another, that the highest value and the essence of man is in the moral dimension. It would be difficult to cite all of them, some names are not even mentioned, although they represent the trend presented here. The views presented here only exemplify the fact that in all periods in the history of philosophy the idea of the primacy of moral dimension was very vivid and espoused by philosophers who were held in high repute.

References

Austin Eugene M. 1935, *The ethics of the Cambridge Platonists*, Philadelphia: University of Pennsylvania.

Bernstein John A. 1980, *Shaftesbury, Rousseau, and Kant*, Rutherford: Fairleigh Dickinson University Press.

Blanshard Brand 1961, *Reason and goodness*, New York: Macmillan.

Bocheński Innocentius M. 1966, *Contemporary European philosophy*, Berkeley: University of California Press .

Bourke Vernon J. 1970, *History of ethics*, vol. 1, Garden City: Image Books.

Cassirer Ernst 1953, *The Platonic Renaissance in England*, Austin: University of Texas Press.

Cassirer Ernst 1961, *Rousseau, Kant, Goethe: Two essays*, Hamden: Archon Books.

Castle Edgar B. 1962, *Educating the good man: Moral education in Christian times*, New York: Collier Books.

Cohen Carl 1991, The case for use of animals in biomedical research, in Baird Robert M., Stuart E. Rosenbaum (eds.), *Animal experimentation: The moral issues*, Buffalo: Prometheus Books, 103-114.

Copeland Frederick 1965, *A history of philosophy*, vol. 7, part 1, Garden City: Image Books.

Drozdek Adam, Beyond infinity: Augustine and Cantor, *Laval théologique et philosophique* 51 (1995), No. 1, 127-140.

Dulles Avery R. 1941, *Princeps Concordiae. Pico della Mirandolla and the scholastic tradition*, Cambridge: Harvard University Press.

Elders Léon 1987, *Autour de Saint Thomas d'Aquin*, v.2, Brugge: Tabor.

Festugière Jean 1941, *La philosophie de l'amour de Marcile Ficin et son influence sur la littérature française au XVIe siècle*, Paris: Librarie Philosophique J. Vrin.

Fischer Kuno 1900, *Geschichte der neuern Philosophie*, vol. 6, Heidelberg: Winter.

Fowler Thomas 1883, *Shaftesbury and Hutcheson*, New York: G.P. Putnam's Sons.

Frings Manfred S. 1965, *Max Scheler: A concise introduction into the world of a great thinker*, Pittsburgh: Duquesne University Press.

Fromm Erich 1956, *The art of loving*, New York: Harper & Row.

Fromm Erich 1947, *Man for himself: An inquiry into the psychology of ethics*, Greenwich: Fawcett 1975.

Garin Eugenio 1965, *Italian humanism: Philosophy and civic life in the Renaissance*, New York: Harper & Row.

Gilson Etienne 1931, *Moral values and the moral life: The ethical theory of St. Thomas Aquinas*, Hamden: The Shoe String Press 1961.

Glathe Alfred B. 1950, *Hume's theory of the passions and of morals*, Berkeley: University of California Press.

Grean Stanley 1967, *Shaftersbury's philosophy of religion and ethics: A study in enthusiasm*, Athens: Ohio University Press.

Irwin Terence 1977, *Plato's moral theory*, Oxford: Clarendon Press.

Kant Immanuel 1977, *Moral and political writings*, New York: The Modern Library.

Kristeller Paul O. 1964, *The philosophy of Marcilio Ficino*, Gloucester: Peter Smith.

Kristeller Paul O. 1972, *The Renaissance concepts of man and other essays*, New York: Harper & Row.

Kuderowicz Zbigniew 1963, *Fichte*, Warsaw: PWN.

Lamprecht Sterling P. 1962, *The moral and political philosophy of John Locke*, New York: Russell & Russell.

Lichtenstein Aharon 1962, *Henry More: The rational theology of a Cambridge Platonist*, Cambridge: Harvard University Press.

Lodge Rupert C. 1928, *Plato's theory of ethics*, New York: Harcourt, Brace.

Mackie John L. 1980, *Hume's moral theory*, London: Routledge & Kegan Paul.

McInerny Ralph 1982, *Ethica Thomistica: The moral philosophy of Thomas Aquinas*, Washington: The Catholic University of America Press.

Nelson Daniel M. 1992, *The priority of prudence: Virtue and natural law in Thomas Aquinas and the implications for modern ethics*, University Park: The Pennsylvania State University.

Petzäll Åke 1937, *Ethics and epistemology in John Locke's Essay concerning human understanding*, Götenborg: Wettergren & Kerbers.

Pine Martin L. 1986, *Pietro Pomponazzi: Radical philosopher of the Renaissance*, Padova: Editrice Antinova.

Przywara Erich 1923, *Religionsbegründung. Max Scheler - J.H. Newman*, Freiburg: Herder.

Raymond Marcel 1958, *Marcile Ficin*, Paris: Les Belles Lettres.

Rousseau Jean-Jacques 1971, *The social contract* and *Discourse on the origin of inequality*, New York: Pocket Books.

Scheler Max 1973, *Formalism in ethics and non-formal ethics of values*, Evanston: Northwestern University Press.

Scheler Max 1960, *On the eternal in man*, New York: Harper.

Scheler Max 1957, *Schrifren aus dem Nachlass*, vol. 1, Bern: Francke Verlag.

Scheler Max 1955, *Vom Umsturz der Werte*, Bern: Francke Verlag.

Scheler Max 1960, *Die Wissensformen und die Gesselschaft*, Bern: Francke Verlag.

Scott William R. 1966, *Francis Hutcheson: His life, teaching and position in the history of philosophy*, New York: Kelley.

Sidgwick Henry 1949, *Outlines of the history of ethics*, London: Macmillan.

Smith Norman K. 1941, *The philosophy of David Hume*, London: Macmillan.

Sorley William R. 1921, *A history of English philosophy*, New York: Putnam's Sons.

Starobinski Jean 1971, *Jean-Jacques Rousseau: La transparence et l'obstacle*, Paris: Gallimard.

Sullivan Roger J. 1989, *Immanuel Kant's moral theory*, Cambridge: Cambridge University Press.

Trinkaus Charles 1970, *In our image and likeness. Humanity and divinity in Italian humanist thought*, Chicago: University of Chicago Press.

Tuveson Ernest 1960, *The imagination as a means of grace: Locke and the aesthetics of romanticism*, Berkeley: University of California Press.

Verbecke Gerard 1976, Man as a 'frontier' according to Aquinas, in Verbecke G., Verhels D. (eds.), *Aquinas and problems of his time*, Louvain: Leuven University Press, 195-223.

Voitle Robert B. 1955, Shaftesbury's moral sense, *Studies in Philology* 52, 17-38.

Willey Basil 1967, *The English moralists*, Garden City: Anchor Books.

Chapter 4

Moral Dimension of Man
and
Psychology

Philosophical ideas presented in the previous chapter are general and not infrequently of the speculative nature. However, it would be interesting to see whether any light can be shed on the problem of moral vs. intellectual dimensions from a scientific angle and whether the idea of predominance of the moral dimension of man can to any degree be substantiated by observation and experiment. Therefore, I would like to turn now to psychology, and in particular, to the domain of psychology which analyzes the problem of moral development.

Moral development research is concerned with such issues as the origin and development of moral judgment and reasoning, the influence of morality on behavior, nature, characteristics and functioning of conscience. Many different theories of moral development were created, some of a speculative nature, and some of them grounded very firmly in empirical research.

There are two paradigms of moral development, relativist and absolutist. Most researchers assume that there is a universal progression of stages in moral development and that "there are universal principles, found at the end of the progression, the cultivation of which will elevate the human condition" (Liebert 1984, 178). Almost two centuries ago Fichte, in the tenth *Address to the German nation*, said that man strives for respect: it starts from an unconditional respect of child for adults and adult, being a measure for himself, seeks respect in those for whom he has respect. This observation founded on common sense and experience was more of theoretical than empirical character.

This paradigm can be found in McDougall and Freud, who

explained morality in a naturalist fashion, and for whom instinctive impulses constitute a core of personality. Man is primarily a part of nature and no socialization changes this fundamental fact. For McDougall, humans have an inborn moral instinct, which develops in a natural way. He distinguished four stages of moral development or "levels of conduct". In the first stage behavior is controlled by instinct, in the second, behavior is still instinctive, but it also is controlled by rewards and punishments, in the third, by the anticipation of rewards and punishments, and in the fourth, an ideal of conduct is the main factor regulating behavior (1918, 186).

For Freud man is composed of *id*, the biological drives or instincts, *superego*, an internalized external authority, and *ego*, the intermediary between *id*, *superego* and the world. *Superego* "represents more than anything the cultural past" (1940, 63); it is simply conscience, that is a set of rules imposed by social environment, which allows man to be a part of society at the cost of suppressing his true self, the reservoir of drives. If the *superego* becomes too powerful, a person becomes neurotic since clashes between *ego* and *superego* lead to the feeling of unbearable guilt. This also is an unwelcome consequence of the progress of civilization that man will be more and more unhappy and trodden by the *superego*: "if civilization is a necessary course of development from the family to humanity as a whole, then ... it is inextricably bound up with it an increase of the sense of guilt, which will perhaps reach heights that the individual finds hard to tolerate" (1930, 89). Suppression of the biological drives becomes intolerable and man can live humanly only if the animal in man is treated tenderly and with lenience. The development of civilization brings more suffering to man, Freud seems to say, since the libidal instincts become unnecessarily tamed. Man can be man if drives are loose, and *superego*'s attempts to make man less animalistic bring about unhappiness, neurosis, and inhumanness. *Superego* is thus an enemy, a necessary evil which should be tolerated in the most curtailed version possible, because by its demands, man loses his humanity. We should submit ourselves to *id* rather than to *superego*, otherwise we slip into insanity. Animality, therefore, is equated with humanity, instincts win over morality, libido gains an upper hand over the rest of personality.

It has to be remarked, however, that such a limited vision of the human being does not have to be necessarily a flaw of psychoanalysis as a whole, but certainly it is a flaw of its version due to Freud.

Freud's merit is earned by introducing many useful concepts (or rather reintroducing them, since the concept of *superego* is due to Heinroth's *Überuns*, *id* to Groddeck's *Es*, and the concept of the unconscious was "topical around 1800 and fashionable around 1870-1880 ... It cannot be disputed that by 1870-1880 the general conception of the unconscious mind was a European commonplace, and that many special applications of this general idea had been vigorously discussed for several decades" (Whyte 1960, 169-70)). But his theory drives man back to the animal level by scrupulously singling out what aspects are animal in man. If human elements are introduced in this theory, they are quickly reduced to this level and treated as basically evils hindering the development of drives. Libido is seen everywhere and the development of civilization is a cosmic error, since it reduces its scope and ultimately leads to neuroses.

One version of psychoanalysis which attempts to overcome such problems is due to Hobart Mowrer. He first observes that if neurosis were caused by *superego* dominated *ego*, that person should be perfectly fitted to the society; but the opposite is true. He attempts to put the simplistic biologism of Freud right side up by observing that neurosis is caused by an *id*-dominated *ego* and its separation from the *superego*. "Instead of seeing conscience as irrational and tyrannical, one now sees the conscience of a neurotic person as trying to 'reach,' 'redeem,' save its owner, and bring him back into community and restore him to fully human status and functioning" (Mowrer 1967, 354). Neurosis is a result of *ego*'s rejection of the voice of conscience and submitting itself to *id*, to the animal and the libidal. Human status can be regained if the weakness of conscience is overcome and *ego* again cooperates with it, *ego* and *superego* become unified and congruent.

Mowrer's contention has been tested experimentally by showing, for instance, that "people who seek psychotherapy for neurotic complaints ... have violated moral laws more frequently than normal people coming from the same socioeconomic background" (Swensen 1962, 372).

Mowrer gives a more humane tint to psychoanalysis by simply showing that humans become neurotic if the animal becomes a dominant element and not the *superego*, the voice of conscience. Restoration of this human element in man is the way of pulling man from neurosis and bringing him back to the fully human status. Therefore, the moral dimension is given dominance in human personality and the attempt of the animal to take *superego*'s position is the beginning of the end, unless the process is reversed.

Mowrer restored the proper proportion between the animal and human in person, but it still did not restore the ethical dimension to its proper form. Mowrer still retains Freud's concept of *superego*. Freud himself was not altogether clear about the meaning of this concept, since sometimes the *superego* was equated with conscience (1940, 62; *The ego and the id* (1923)), sometimes conscience was one of the functions of *superego*, however the other function was self-observation, "an essential preliminary to the judging activity of conscience" (1933, 60). This faculty is imposed on man, who after birth is just a bundle of "instinctual drives." This identification of conscience and *superego* is sometimes found unsatisfactory and it is insisted that these two faculties should be clearly separated.

For John W. Glaser, *superego* functions on a primitive levels of life, such as toilet training, and even "higher animals are said to have a *superego* when they have been trained." On the other hand, values concerning life as a whole and relations with people are a concern of conscience. Conscience is "the call of genuine value", "an insight into love, the call issued by the ultimate value and promise of love." This element is the most important, since "the moral action of man is love" (Glaser 1973, 168-9, 173-4). What Glaser finds unsatisfactory is almost a biological level to which conscience is reduced. *Superego* does not really help a person to break out of narcissism, to end with selfishness. Rules are obeyed on account of *ego*, not on account of others. *Superego* watches over drives which are truly the driving force of *ego*. Goodness is a pragmatic concept used in terms of self-interest and in the interest of *id*'s drives.

Therefore, in Mowrer's conception of person, this higher level of moral dimension is missing, psychoanalysis, albeit right side up, is still stunted. It is due to the fact that psychoanalysis is largely a naturalist view, in which man is a part of nature and remains bound to the natural, even animal level, and both moral and cognitive dimensions are secretions of the natural mechanisms which merely serve adaptation and survival. Natural impulses are the driving force of human life, possibly transformed to allow social life, and moderated or suppressed to sustain an appearance of supra-animality. Biological ties of humans to nature are upheld in all domains and dimensions, biological factors remain salient and psychology turns into a branch of biology.

This biological approach had a very strong influence on behaviorism which became a leading paradigm in psychology, especially in 20s, 30s, and 40s. This paradigm offered seemingly more scientific approach to psychological phenomena, which eventually led to a complete disinterest in the *psyche*. What was important in this was the stress put on empirical data to the extent of disregarding everything what cannot have directly have tied to the empirical, or inferred from the data.

Empirical studies of morality began at the end of the 19th century, and one of the first studies was conducted by Osborne (1894) whose questionnaire was designed to determine the "ethical content of children's mind." His conclusion was that truthfulness and obedience were the principal features of children's ethics. An early study of moral system of college students was conducted by Sharp (1898). Later research was mainly based on vocabulary tests, tests involving comparison of acts, cross-out (X-O) tests, etc. (Pittel, Mendelsohn 1966, 80).

One of the first empirical contributions to the problem of morality was Macauley and Watkins's paper (1925). They simply asked over 3,000 schoolchildren to make, for instance, a list of "the most wicked things anyone could do," and to choose a person they wanted to be like and arguments for this choice. One conclusion was the statement that children develop their value system by accepting social norms, but it is also possible to determine some stages of moral development (Kay 1969, 34).

May and Hartshorne directed a five year project. They emphasized behavioral measures and attempted to evaluate the connection between moral knowledge and behavior; the correlation found was .5, thus moderate. Their findings seemed to indicate that situational factors have a major influence on cheating or resistance to cheating. Situation was determined by degree of control, service etc. Correlations between cheating and some situations ranged from zero to .45. Thus, "moral conduct is in large part the result of an individual decision in a specific moral conflict situation" (Kohlberg 1964, 448). Moral knowledge was highly correlated with measure of intelligence (.7), but insignificantly with the measures of behavior, inhibition, cooperation (.25) (Pittel, Mendelsohn 1966, 83). It was one of the largest research project at that time, including thousands of subjects, however, very little pattern was found. Children cheated in one situation, and were honest in another, and their conduct was barely related to what they knew about proper rules of conduct. But maybe letting other's copy was helping a friend. Thus children's acts have to be understood in the context of a child's world. As Kay points out, May and Hartshorne investigated particular moral traits, but their project was concerned with morality. They emphasized particular elements of moral behavior and simply generalized for the whole of morality: *pars pro toto.*

Morality develops, but moral conduct can be molded by situation. May and Hartshorne did not see, like Macaulay and Watkins, any regularities in moral development. But there does not have to be a contradiction between seeing that actions are determined by situation, and seeing some general pattern. Consistency is not a property of conduct, but of our assessment of conduct. The same conduct can be judged to be consistent by one and to be lacking consistency by another. Thus, according to children's value system, no inconsistency may be seen in what, for adults, is plagued by such. In the perception of children, honesty may not be important at all in some situations, or is overridden by other concerns, such as helping a friend. Hence, what appears to be inconsistent is perfectly justifiable by the standards of the children's value system.

A major break with the behavioristic approach is the moral-reasoning orientation of Jean Piaget, who pointed to the importance of cognitive elements in morality, such as knowledge of moral rules, anticipating consequences of moral decisions, or ways of compensating for irresponsible actions. Piaget stressed measures of judgment and wanted to examine systems of moral judgment and their relation to action, although, unlike behaviorists, Piaget was mainly interested in what children think about rules, including moral rules, and not in what they do.

Piaget conducted experiments with some one hundred Swiss preadolescent children (age 6 to 12), with a game of marbles and the children's perception of its rules, and with moral assessment of stories read to them. He described his results in *The moral judgment of the child* (1929). The game of marbles, although it does not have by itself a moral significance, is played in accordance to certain rules and Piaget wished to observe the way children treat these rules and observe them. This would give him a way of judging how morality develops in children, since "all morality consists in a system of rules, and the essence of all morality is to be sought for in the respect which the individual acquires for these rules" (Piaget 1929, 13).

Piaget found in children two different forms of moral judgment, which occur in a successive order. The first, called heteronomous, was based on the material consequences of wrong-doing. This type of morality leads to moral realism which consists in unquestioned obedience to rules. It is a morality of obedience which does not take into account intentions nor relations and demands that "the letter rather than the spirit of the law shall be preserved" (p. 111). This results in objective responsibility which judges offenses according to the degree of violating laws rather than to motives and absence or presence of premeditation. Therefore, "veracity is external to the personality of the subject and ... lie appears to be serious not to the degree that it corresponds to the intent to deceive, but to the degree that it differs materially from the truth" (Piaget, Inhelder 1951, 126). In this light, it is a bigger lie to say that one ate ten pounds of chocolate than to say that one did not witness a friend shoplifting which he actually did see.

The stage of moral realism is inevitable regardless of the atmosphere at home; it can be more evident when parents espouse a rigorous approach to child education, but it still can be found even in a very lenient environment. Piaget himself quotes an example of his daughter, Jacqueline, then two years old, who was never punished, nor given duties, nor required obedience without previous discussion. But still, she cries and "shows signs of remorse" when she is unable to finish her usual meal because she does not feel well. The rule, even the law of (her) universe, says that meal has to be had at such-and-such hour, hence, if broken, the feeling of guilt arises, because a misdeed has been done. Consolation on the part of her mother made the remorse smaller, but not inexistent (Piaget 1929, 178, 181).

For Piaget, moral realism is a form of general realism, characterizing the whole of child's thinking. He described this process at length in *The child's conception of the world* (1927). In this work he poses the question of whether a child makes a distinction between internal world and objective reality, or whether and to what extent a child is egocentric (implying no negative connotations). This problem is divided into three questions: 1. the question of child's realism - whether a distinction between subjective phenomena and the reality is perceived, 2. the question of animism - to what extent animate characteristics are attributed to objects, and 3. artificialism - the problem of a division between man-made artifices and natural objects. Piaget analyzed this problem in all generality, but these phenomena are also applicable to moral laws. Hence, a child, being a moralist, reifies moral rules as well. Interestingly, up until the age of 7-8 a child "always regards the notion of law as simultaneously moral and physical ... there does not exist for the child a single purely mechanical law." Clouds move to bring rain, sun shines to mark the daytime, and it all happens "because things have to be so in virtue of the World-Order. In short, the universe is permeated with moral rules; physical regularity is not dissociated from moral obligation and social rule" (Piaget 1929, 188-9). Child is a perfect holist, panpsychist and teleologist, seeing the world as a grand living organism making nothing in vain, filled with physical regularities that have moral significance. Motives in violating these regularities have no significance, and as one is burned by putting his hand into fire, regardless of his intentions, so one should be punished when breaking a moral law - regardless of motives.

The first period, ended at 7-8 years of age, is followed by an intermediate period of age between 8-11 and characterized by "progressive equalitarianism" when the morality of obedience and constraint is being replaced by the morality of cooperation, or reciprocity (Kay 1969, 41-2). During this phase "rules and commands are interiorized and generalized" (Piaget 1929, 195, 315).

In the second, autonomous moral stage starting at the age of 11-12, justice becomes the most important factor as a result of mutual respect and reciprocity among peers. Justice becomes more important than obedience, and plays the same role in the affective realm as the norm of coherence with respect to cognitive operations. In the heteronomous stage, the severity of moral damage is judged by the amount of material damage: the more cups broken, the more serious is the offense. In the autonomous stage, it is the intention of the offender that constitutes the mark of offense. The ethics of obedience turns into the ethics of mutual respect, which is based on good, and not duty. Then the idea of justice emerges, which is not based upon authority, but springs form the sense of cooperation and of mutual respect. This all leads to complete autonomy of the child. And autonomy appears "when mutual respect is strong enough to make the individual feel from within the desire to treat others as he himself would wish to be treated" (Piaget 1929, 196).

This succession of stages is general and although it is "a function of the mental age of the child" (p. 313), it is not inevitable. Child is not a close microcosm and "mental age" cannot bring a child by itself from the heteronomous ethics to autonomy. The child is a part of society and influences of this society play a pivotal role in child's moral development. First, "neither logical nor moral norms are innate", therefore, the norms themselves have to come from the outside, and Piaget seconds the sociological theory of Durkheim in that respect, according to which "society is the only source of morality" and "morality presupposes the existence of rules which transcend the individual and these rules could only develop through contact with other people" (pp. 398, 327, 344). Also, continuous authoritarian pressure exercised by adults on children inhibits both intellectual and moral development. In order to bring child to the level of full autonomy, cooperation is necessary. Although it initially leads to criticism and individualism on the part of the child, it eventually brings the desired autonomy and full development of personality. In this process, adults should be first of all collaborators not attempting to impose onto children a system of norms, since "the social life of children amongst themselves is sufficiently developed to give rise to a discipline infinitely nearer to the inner submission which is the mark of adult morality" (p. 404).

Piaget stresses a parallelism in the development of intelligence and of moral dimension. Logic to him is a set of rules of control utilized by intelligence, and "morality plays a similar role with regard to the affective life" which is summarized by a statement that "logic is the morality of thought just as morality is the logic of action" (pp. 400, 398). In *The moral judgment of the child*, Piaget makes few remarks concerning this issue, but this discussion is continued in his lecture notes, *Intelligence and affectivity* (1954). In this book he argues for the impossibility to separate in real life the cognitive and affective elements. Both are present in all actions, one can be separated from another only in theory, and "there is a constant and dialectic interaction between affectivity and intelligence" (Piaget 1954, 25). Affectivity is an "energy source" on which functioning of intelligence depends, but not its structure.

Although a system of values can be detected in children before the age of two, it is with the appearance of language that feelings become stable and durable. "This ability to conserve feelings makes inter-personal and moral feelings possible and allows the latter to be organized into normative scales of values" (p. 44). However, it is not feeling as such which is conserved, but a certain scheme which allows constructing and reconstructing it. Feelings, as Piaget stresses, have no structure, therefore, there does not exist a scheme of feelings.

Instead, there are schemes which have to do with objects and with people, the latter, as a result of "intellectualization of the affective aspect of our exchanges with people", only in this sense can be called affective structures (p. 73).

Piaget tries to do his best to keep balance between intellect and affection. On the one hand, he stresses a profound effect affectivity can have on development of intelligence: it can influence its functioning and content. But it cannot have an impact on the structure of intelligence, and this probably allows him to say that there is "no modifying action" of affectivity on intelligence, or vice versa, and that such a view is unintelligible. They are heterogeneous domains, and this fact does not square with a claim that affectivity can modify contents of intelligence, or with a statement that values play "a distinct role in primary actions and are evident from the moment the subject begins to relate to the external world" which starts at six months of age (Piaget 1954, 32, 73-4). Piaget seems to be willing to see only parallelism between intellect and affectivity, whereby the problem of mutual influence between these two domains seemingly disappears. On the other hand, he stresses very strongly that intellectual and affective elements are present in all behavior so that behavior should be classified rather than different powers of human spirit.

In spite of his hesitance on this issue, Piaget seems to ascribe a formative role to both intelligence and affectivity and sees mutual modifying influences. The influences of affectivity on intelligence have been already mentioned. Intelligence, on its part, molds affectivity, since "affective structures ... result from intellectualisation" which "exists from the moment feelings are structured" (p. 9). Affectivity, as devoid of structure, can be structured by something which has such a structure, that is, intelligence, whereby "affective structures become the cognitive aspect of relationship with other people" (p. 74). It is true that child manifests in his behavior "sympathetic tendencies and affective reactions in which one can easily see the raw material of all subsequent moral behavior", however these tendencies and reactions are moral only if they are subjected to rules, that is, if they are structured (Piaget 1929, 398). Otherwise, they are premoral sentiments which bring a child, an amoral being, to the threshold of morality. And in this process, intelligence has a primary role to play; it builds and properly forms affectivity which is the source of energy of intellectual powers. Affectivity uses intelligence to achieve this goal, fueling it all throughout its operation. Intelligence disappears without affectivity,

affectivity without intelligence cannot make man what he is: a moral being.

The latter conclusion might not be acceptable to Piaget. In fact, his interest in morality was rather casual in his research and writing. Piaget, a structuralist, was mainly interested with orderly and structured characteristics of man, and therefore he concentrated primarily on cognitive dimension which offered more order than moral dimension. He seems to have made his sketchy lectures on affectivity vs. intelligence only reluctantly to avoid an accusation of over-intellectualist bias of his psychology and that his "study of intellectual development lapsed into intellectualism" (Piaget 1954, 1). This opinion was not entirely unjustified as it may be seen in his confession, when he says "I have no interest whatsoever in the individual. I am very interested in general mechanisms, intelligence, and cognitive functions" (Piaget, 1971, 211). From here stems his emphasis on that morality can be properly discussed only if rules are in place, whereby he purges it from its specificity; morality concerns what is wrong and good more than it concerns rules. The latter is used to express the former, not the opposite. In this way Piaget "implies that children develop the idea of a rule and derive notions of right and wrong ... from this idea" (Sugarman 1987, 89). This fact is not taken into account by Piaget, and his definition of morality by the concept of good seems to be very indirect and oblique (e.g., Piaget 1929, 195).

Study on morality was not a centerpiece of Piaget's vast research and it is certainly an overstatement to say that "*Moral judgment* is central to understanding the appeal of Piaget to psychology" (Cohen 1983, 62). Piaget conducted his research as a consequence of his interest of structured dimension of human mind and wrote only *Moral judgment* on this subject at the early stage of his scientific work. He mentions moral judgment in children in later works only occasionally, hardly creating an impression that he refers to something "central to understanding the appeal of Piaget to psychology". Nevertheless, Piaget is very important for the psychology of moral development and his contribution is here very significant.

It is Piaget's merit that he recognized a different type of morality beyond what was recognized in his day. Most moralists saw morality as a result of imposition and restraint. Piaget has shown that it is the initial step only, the second step being autonomous morality developed by cooperation. The first stage is necessary to form all necessary means enabling child to become a member of society. On both intellectual and moral levels it leads to acquiring means of stability in expressing thoughts and feelings, to consistency of theoretical and practical reasoning, and to uniform view of the world and society. Without it, autonomy would not be achievable, without a stage of intellectual and moral submission a child's personality would be crippled and underdeveloped. Social life has to enforce something onto children first in order to enable him to live social, but also autonomous, life.

It is also Piaget's merit to see parallelism between intellectual and moral developments. Morality, regardless of a limited interpretation Piaget ascribed to it, is developing in a child alongside intelligence, the rate of moral and intellectual growth is correlated, and, most of all, is always the same for all children. It is a general trend, Piaget maintained, and in spite of some possible variations in the rate of this growth, it is a universal tendency which can be traced in children of all social strata of all cultures. When contrasted with a behaviorist view that morality is just a result of training and conditioning of a child who accepts rules imposed by the adult, or with psychoanalysis which propounded an overwhelming *superego* as the repository of moral rules, Piaget's conception was quite revolutionary. There is a universal order in the development of moral understanding of rules, justice, development in perception of the role of the self in society, if only a microsociety of children. Regardless of how morally developed is an immediate social environment, a child has to pass through the stage of egoism and moral realism. Regardless of how bad is this environment, children reach the phase of moral autonomy, of moral cooperation, subjective responsibility, and distributive justice.

Piaget's work on moral development, being just a side-issue in his work, remained unfinished. He stressed mainly the intellectual aspects of moral development. He also mentioned other factors; cooperation, for example has a great weight in moral development (Piaget 1929, 35, 187), but social and familial influences appear to be merely a side-issue in Piaget's theory. Also, although Piaget was interested mainly in child psychology, "there is the odd feeling that, for him, the pinnacle of development was reached around the age of 12 to 13" (Cohen 1983, 64).

Piaget's work was continued by many researchers. B.H. Smith distinguished three stages in moral development, W.K. Brennan four stages, while R. Goldman applied Piaget's scheme to religious thinking. In 1967 R. Loughran replicated his research with adolescent children, fully confirming Piaget's findings (Kay 1969, 49). Also, Leonore Boehm demonstrated that the stages found in Swiss children by Piaget can also be found in American children. She found that they go through these stages much faster, which is a result of different upbringing practices in these countries. American children become mature faster, since parents want them to become independent, self-confident, and accepted by others. They become exposed at an earlier age to the scrutiny of peer-group as the main source of acceptance; parents treat them as equals, children are emancipated early from their parents. European children are longer dependent on parents, parents are the main authority to them. The conscience content differs also. "Whereas the American child's conscience is turned, primarily, toward social adjustment, the Swiss child's is geared toward character improvement" (Boehm 1957, 207).

But Piaget's work found the most comprehensive continuation in Lawrence Kohlberg.

For Kohlberg, human development, a basis of moral development, is a consequence of an interaction between an organism and environment, and not of direct reflection of one on another. Cognitive-structural component organizes, but does not control human development, hence it is not sufficient for this development. According to cognitive-developmental position, there exist cognitive stages detectable in the life of each human. These stages are qualitatively different, but they form an invariant sequence and constitute structured wholes, and they are hierarchically integrated, i.e., operations proper to lower level do not entirely disappear, but are integrated at a higher stage to be used as a last resort (Kohlberg 1984, 14). This doctrine of different stages implies that cognitive development is not a *direct* result of parental teaching, social situation, education, etc.

Associated, and in fact, intertwined with cognitive development is moral development of man, which, although correlated with cognitive development, is slightly behind it (Kohlberg, Gilligan, 1971, 1071-5). At the same time, neither cognitive nor moral development is of purely biological nature. "It seems obvious that moral stages must primarily be the product of the child's interaction with others, rather than the direct unfolding of biological or neurological structures" (Kohlberg 1964, 456). In this Kohlberg differs from Piaget, for whom development of cognitive development was of biological origin. "For Piaget, the one-time biologist, intelligence can be meaningfully considered only as an extension of certain fundamental biological characteristics," and its functions are "a special form of biological activity" (Flavell 1963, 41-2).

In his 1958 doctoral dissertation Kohlberg studied moral development and reasoning of 72 boys of ages ten to sixteen, using Piagetian approach. "Kohlberg (1958) began his work with a consideration of conduct when he selected delinquents, assuming that through understanding of moral development he might gain knowledge of how to increase their level of morality" (Burton 1984, 199). He distinguished six stages of moral judgment, which later were only slightly modified.

Level I. Premoral or preconventional.

Stage 1. Punishment and obedience orientation. Motives are disregarded, acts are judged by the size of physical damage.

Stage 2. Naive instrumental hedonism. Acts are acceptable if they serve our needs and they do not deprive others of meeting their needs.

Level II. Morality of conventional role-conformity.

Stage 3. Good-boy morality of maintaining good relations, approval of others. Acts are judged by altruistic or selfish motives; circumstances are used to excuse an act.

Stage 4. Authority maintaining morality. Acts are judged wrong if they violate a rule and harm others, regardless of motives and circumstance.

Level III. Morality of self-accepted moral principles.

Stage 5. Morality of contract, of individual rights, and democratically accepted law. A general rule is that the end does not justify the means.

Stage 6. Morality of individual principles of conscience. Moral principles are the unmistakable guides of actions.

(Kohlberg 1964, 461; 1984, 49-51; for six different versions of definitions of these stages presented by Kohlberg see Bergling 1981, 27-42).

Morality of preconventional level characterizes children under 9, some adolescents, and criminal offenders. Adolescents and adults are usually in stages 3 and 4, and only some adults after age 20 attain stages 5 and 6. Persons at preconventional level do not fully comprehend what is commonly accepted in a given society, which comes on the conventional level, when individuals abide by rules held in society and by expectations conventionally accepted. The highest, postconventional level is reached when a deeper understanding and better insight of such rules is attained and the underlying principles

become more important than they conventional, socially accepted formulations. For instance, a person on preconventional level does not
steal out of fear of being caught: law punishes perpetrators, therefore,
I do not steal, since I can be harmed by this law. A person on the conventional level justifies not stealing by the good of others, the good of
neighbors, community, or society as a whole: I do not act unlawfully,
since, I can harm others, or I can damage the tissue of the social
bonds. A principled person does not steal, since it is wrong, since the
rights of others are trampled upon. A moral principle of respect to
others takes precedence over acting according to the law, and the law
is viewed as valid and acceptable only if it is firmly based upon such
principles.   Moral considerations are more important than legal
principles (Kohlberg 1984, 172-3, 177-9).

      In Kohlberg's theory, attaining a certain level of logical operations
is needed for achieving a high level of moral development, since
principled moral reasoning requires development of reasoning as such.
Tomlinson-Keasy and Keasy (1974) showed that it is the fact both for
girls and boys; cognitive growth is a necessary condition for moral
development.   Subjects on the lowest (operational) level of logical
thinking were characterized predominantly by preconventional moral
thinking, but with an increase of their formal operation capacity they
usually advanced to conventional and principled moral thinking.
However, logical reasoning alone does not account for moral development, since subjects on the same level of logical thinking did not
necessarily belong to the same stage of moral development. Kohlberg
found that moral judgment is moderately correlated to IQ, $r = .31$, quite
highly related to age, $r = .59$, and although the level of moral thought
does not entirely coincide with intellectual level, "a more mature level
is a more moral level" (Kohlberg 1974, 465; 1984, 171). Studies with
mathematically gifted students indicate that they are more mature than
students in general. "They are more responsible, more dependable,
more perspicacious in their dealings with the whole structure of their
environment, and more likely to take firm and upright stance regarding
moral matters" (Weiss, Haier, Keating 1974, 136).

      These findings are corroborated by an overwhelming evidence that
among children, IQ and cheating are negatively related; however, it
does not mean that more intelligent kids are more honest (Liebert 1984,
188). Cognitive development forms a foundation for development of
intentionality in moral thinking, which overrides consideration of
physical consequences of certain acts as the only criterion of moral

assessment. Every culture, social class, or nationality studies (including Chinese, European, Indian or Mexican) showed a growing tendency to including intentions in moral reasoning which was invariably correlated with mental development (Kohlberg 1984, 42).

However, high intelligence does not ensure high morality and affection, and "one may be a theoretical physicist and yet not make moral judgments at the principled level" (Kohlberg 1984, 64). Let us recall that both Joseph Goebbels and Pol Pot had PhDs. Interestingly, correlation between moral maturity and IQ is larger for below-average intelligence ($r = .53$), but insignificant for high IQ ($r = .16$). In other words, high intelligence does not ascertain high moral standards. What it does is accelerating the process of moral development, but lower intelligence does not prevent from attaining even a principled level, although it can take longer. "IQ is then a better indicator of early rate of development than it is of terminal status, which is more determined by social experience" (Kohlberg 1984, 65). Sometimes even this statement may seem to be too generous, since, as shown by Douglas Heath's research, many times a scholastic aptitude test of men under thirty "was inversely related in this group to many measures of their adult psychological maturity, as well as of their judged interpersonal competence", which includes such morally-oriented characteristics as compassion and commitment to the democratic values (Heath 1977, 177-8).

However, another possibility is not considered in cognitive-developmental approach that namely cognitive skills are accelerated by person's morality. Getzels and Jackson (1962, ch. 4) conducted a study with gifted children, by dividing them in two groups, moral group and adjusted group (which more or less corresponded to Kohlberg's conventional and principled levels), and analyzed their ratings in different cognitive skills. "The results are quite provocative." Although intelligence scores were almost the same, verbal and numerical achievement was much better in the moral group. The authors mention as an explanation the mere possibility that "the moral students take school more seriously and are more diligent about applying themselves to their lessons." That is, their sense of moral obligation and responsibility causes them to apply themselves to learning more than others, whereby cognitive level can be increased and they can become more efficient and more useful in their future actions. Intelligence is largely ineffective if not amplified by knowledge, gaining which does not come effortlessly. Responsible children are able to overcome possible inconveniences, stresses, and a temptation to take an easy way out, by taking learning, expanding their knowledge and cognitive skills seriously. Moral dimension induces development of cognitive dimension, so that the latter can be more effective in service of the former.

All this indicates that reliance on reasoning in moral matters proves to be a very efficient technique, but insufficient if not used in conjunction with emotional imposition and empathy-oriented methods. Intelligence is a necessary substratum for morality, and men who would end their development with a high level of cognitive skills only would not be fully human. On the other hand, morality not supported by cognitive skills would be without effect, therefore it influences its growth.

The stages of moral development to be considered stages should form an invariant sequence, i.e., they should unfold in one, upward direction in moral development of an individual. Bandura and McDonald (1963) were first to investigate the problem of learning to reason according to the principles of a lower and a higher stages than the stage characterizing the subject. The conjecture was that children will not accept moral reasoning too far from their level. Elliot Turiel conducted an experiment with 44 seventh-graders to validate that conjecture. First, he determined the level of each subject by their responses to six Kohlberg's stories used for such tests. Next, he exposed them to moral reasoning different from their own stage by reading them three stories and instructing them to take the role of the main character. Experimenter played the role of an advisor presenting arguments of both sides of the conflict. The arguments were presented in form proper to one stage below, one stage above, and two stages above the subject's level. Last, subjects were tested again on all nine stories to determine their learning and understanding. Turiel found that next stage was assimilated easier than the previous stage even if concepts proper to this next stage were understood worse than concepts of the preceding stage. It also substantiates the view that the development consists in "reorganization of preceding modes of thought, with an integration of each previous stage" rather than adding new elements (Turiel 1966, 316). Rest, Turiel, and Kohlberg (1969) replicated Turiel's study and found that assimilation of thinking from the stage above is made more readily than from stage below or two stages above. Tendency to choose advice from a stage above was dominant.

An upward movement in moral development was found among

Turkish children between the ages 4 and 16, among whom Ugurel-Semin (1952) investigated a pattern of sharing. It turned out that 14% of almost 300 subjects were selfish by keeping more to themselves and giving less to others, 42% were equalitarian by dividing equally, and 44% were generous by giving more than retaining to themselves. The selfish tendency was strongest between the ages 4 and 6 and disappeared after age of 12. Sex differences did not influence this pattern. Children from large families were more generous. These findings were confirmed by Handlon and Gross (1959).

Another way of testing the hypothesis concerning an upward movement in moral development is longitudinal studies. Several such studies have been conducted. For example, Kohlberg and Elfenbein (1975) for 20 years studied 30 males, testing them in three year intervals, and they found very strong support for progressive changes in moral development. Kohlberg and Nisan reported the results of studying 23 males, age 10-28 from a Turkish village, and two cities, some of them for over 12 years. There was a clear indication of stage progression: out of 35 changes (passing from stage to stage) only four were recessive, no skipping of stage occurred. The rate of moral development was slower in the village than in the city (Kohlberg 1984, ch. 8). Snarey, Reimer, and Kohlberg also reported results of a longitudinal study of 64 male and female adults from an Israeli kibbutz (Kohlberg 1984, ch. 9). Generally, a progressive sequence was found from stage 2 to 4/5 between ages 12 and 26. No subject skipped a stage. However, results of four other longitudinal studies showed that progression is significant among children; one study reported progression among adolescents and one rejected it, two studies rejected progression among adults (Bergling 1981, 58). These results indicate that for stages 1-4 the development is progressive, and for stages 5-6 a regression may be expected, more frequently anyhow than for stages 1-4.

Another assumption stage theory makes is a universalist claim that regardless of country or nationality people go through the same sequence of stages in their moral development.

Studies have been conducted in various countries and the proof of the universality thesis does not seem to be consistent. In Canada, Great Britain, Israel, Taiwan, and Nigeria all six stages have been found, in Bahamas only three, in British Columbia only two, and in Turkey five (Bergling 1981, 63-4). John Snarey (1985) reviewed forty-five studies in twenty-seven countries all supporting Kohlberg's universalist claim: "The invariant sequence proposition was found to be well supported, because stage skipping and stage regression were rare and always below the level that could be attributed to measurement error" (p. 226).

The contention concerning invariability of the stage sequence is frequently challenged, especially in the face of delinquency. Interesting in this respect are Kohlberg's findings concerning perception and interpretation of dreams. He found that adult members of a Malaysian tribe of Formosa, the Atayal, believe in the reality of dreams, which they equate with ghosts and souls. However, boys until age of 11 develop the same conception of dreams as American children. "Both the youngest child's conception of the dream as real and the school age child's view of the dream as subjective are their own; they are products of the general state of the child's cognitive development rather than learning of adult teachings (though the adolescent's later 'regression' to concepts like those held by the younger children does represent such direct cultural learning)". And according to Turiel, such a regression is not truly regression, since it represents superimposing of content learning upon cognitive structure (Kohlberg 1984, 21,23-4).

Very strong cultural or sub-cultural influences can explain a regression in moral development. While a normal tendency is an upward moral development so that, for instance, college students are never below their high-school level, the only cases of regression found in longitudinal studies are delinquents. Moral judgments of Adolf Eichmann fall into stages 1 and 2, which can be explained by a regression in moral development caused by an intensive Nazi indoctrination during his adolescent years (cf. Kohlberg 1984, 54-5, 60-61).

Kohlberg establishes six formal stages that can be filled with different content. These stages determine the way moral norms are acquired and substantiated, but not what these norms are. It is analogous to language acquisition; there exists in children an ability to acquire *a* language, not *this* or *that* language. Depending on linguistic environment, a child is able to learn language apparently with little effort. This ability disappears in adolescence, and the learning process is different in adolescent or adult than it is in child. However, this ability is not limited to one particular language or a group of languages, but it enables child to learn any language, or, if taught concurrently, two or more languages. This ability is resilient enough to allow acquisition of language from any linguistic family, so that Iranian child born in Washington speaks fluent English with no accent, and a child of American parents born and educated in Teheran speaks perfect Farsi.

Similarly, regardless of culture and nationality man's moral development goes through the same stages, but the content of the stages, particular rules, ethical precepts, moral dos and don'ts differ from one culture to another. However, plasticity of moral stages allows humans to adopt rules to their level of development and to adopt themselves to these rules. Content of rules, moral stages, and moral reasoning varies, their structure and developmental sequence remains the same (cf. Kohlberg 1984, 582). Even if people from different stages hold to the same values (do not steal, do not hit, etc.), their rationale differs. The significance of these values is not the same, and their positions in the system of values and norms may vary significantly. Moreover, the same symptoms of violating rules are interpreted differently. Children may interpret the feeling in the pit of their stomach as a fear of punishment, adults would see in it a physical manifestation of their guilt and conscience pricks (cf. Kohlberg 1984, 66).

This observation is supported by results of research of Israeli adults. Subjects were interviewed several times with regard to various topics of moral significance and in 94% of the cases all reasoning was performed at one level or two adjacent levels, only 6% at three levels, also adjacent. The various moral topics referred to such issues as law, conscience, punishment, contract, and authority. Correlations between scores for these topics were all positive and significant (between .42 and .74), and, most importantly, factor analysis indicated a presence of one factor accounting for correlations between these topics. This suggests that "there is a general dimension of moral reasoning that is not issue specific" (p. 610), that is, stage is an organizing whole of moral reasoning which regardless of a particular topic manifests itself in a similar manner and operates similarly in different areas.

Existence of abilities does not guarantee that they will trigger themselves and that development will unfold regardless of circumstances. Therefore, children would not learn language if they were isolated or prevented from hearing language. Also, there has to be an adequate surrounding (natural, familial, and social) in order for moral development to take place. In an environment not conducive to this development, growth will be unsatisfactory, since the progression of stages may stop prematurely. This seems to be the case with Turkish village in which only two out of sixteen oldest subjects can be seen as belonging to stage 4 in comparison with nine out of twenty for the city subjects. As indicated, subjects in Bahamas and British Columbia did not reach stage 4; also, C.P. Edwards found in Kenya stage 3 to be the final point of development, which all substantiates her supposition that "stage 3 is a necessary and sufficient level of functioning in societies having a social order based on face-to-face relationships and a high level of normative consensus. These conditions do not necessitate differentiation and integration beyond stage 3" (Kohlberg 1984, 593).

Different studies indicate that although stages 1 through 4 can be found in all cultures, stages 5 and 6 have a more restricted application. For example, these stages are absent in two investigated villages of Yucatan and in Turkey. Therefore, "it is possible to view stages 4, 5 and 6 as alternative types of mature response rather than a sequence" (Kohlberg 1984, 57), whereby moral development would go through stages 1-4 only, the fourth stage having three variants. Lutz Eckensberger (Locke 1985, 23), suggested that stages 4/5, 5, and 6 are more sophisticated versions of stages 1, 2, and 3, which may point out that the former stages can still be seen as separate stages. C. Holstein would retain stage 5, but she sees stages 3 and 4 as parallel stages specific for males and females (Bergling 1981, 81). Kohlberg does not consider seriously the possibility he himself suggested, by seeing in stages 1-6 self-standing steps in moral development. The main problem, however, is with the lack of empirical support for the claim of universality of stage 6.

Kohlberg does not reject the sixth stage from his theory, but he gives it a different significance. Stages 1-5 are hard stages, that is, they conform to Piagetian definition of stage, they meet formal criteria and empirical studies show their generality and sequentiality. On the other hand, stage 6 is a soft stage, i.e., it includes also affective and reflective characteristics, and as such it does not entirely fit the framework of cognitive-developmental approach of Piaget and Kohlberg (Kohlberg 1984, 237).

Stage 6 has been introduced by Kohlberg in 1958, and its existence is based on "the writings of a small elite sample, elite in the sense of its formal philosophical training and ... its ability and commitment for moral leadership". In research conducted on large samples of subjects, stage 6 was virtually absent. However, Kohlberg insists that this stage should not be rejected. In giving a reason for this decision, Kohlberg refers to methodological and philosophical discussions concerning scientific theories. There are different theories concerning what is scientific method, in particular, observation, experiment, formulation, and testing of hypotheses, deductive vs. inductive reasoning, models of theory formation, of explanation and prediction, etc. Also, "although scientists do not actually agree on many scientific conclusions, the ideal of agreement is still central to science and scientific development. Similarly, a rational ideal of moral development implies the need for *moral agreement* in conclusions about moral problems" (Kohlberg 1984, 272). That is, like scientific development leads, or should lead, to a unifying vision of the universe and a unifying vision of scientific method, so moral development should reach a point which would be acceptable to all. It is said that stage 6 principles are not universally applicable but universally acceptable, "acceptable to everyone who reasons properly about them" (Locke 1985, 24). However, it seems that for Kohlberg they should also be universally applicable, at least in the long run.

At other stages conflicts are possible (although not necessarily inevitable), so that conflicting parties can represent the same stage of moral development, although particular principles and views are in conflict. But as in science, an underlying belief is that there exists only one scientific truth, so in ethics an underlying conviction may be that there exists one good, or - as Kohlberg might phrase it - one ideal of justice, which is attainable if only by few. With such a conviction Kohlberg endorses a Platonic tradition of immutable ideas which can, at least in principle, be known, a tradition elaborated for ethics by Max Scheler. It is also a tradition of teleological development of the world, or at least of some of its part, which attained its grandest expression in Hegel, for whom the *Weltgeist* goes through certain stages dialectically to attain self-knowledge, and in Teilhard de Chardin, who saw in the Omega Point the goal of evolutionary development of the world. Admittedly, stage 6 is mostly a theoretical construct, although there are some empirical studies indicating that stage 6 subjects were able to agree on moral dilemmas on which stage 5 subjects disagreed (Kohlberg 1984, 273).

Even if stage 6 is accepted as a real phase in moral development, there still remains a problem of convincing others about the validity of our own assumptions. Is an argument possible which is based solely upon moral assumption? If yes, then morality would justify itself, which may not be convincing. We can refer here to Hans Gadamer's concept of the hermeneutic circle from which it is impossible to break out to substantiate systems assumptions. Paul Ricoeur says that this circle "constitutes a vicious circle only for analytic understanding, not

for practical reason" (after Carter 1985, 14), which means a priority of
practical reason, as in Kant. Kohlberg's answer to this problem has a
Kantian spirit. To answer the question, "Why be moral?" Kant himself
referred to immortality of the soul, freedom of man, and existence of
God as postulates of practical reason substantiating claims of this reason
and endowing moral principles with meaning. Kohlberg introduces in
this context stage 7, "a high soft stage", that has a religious orientation
in distinction to other stages which lack such necessary connection with
religion. This stage justifies, in fact, all other stages, since it places
assumptions of moral nature on a more solid, or more general foundation which is a vision of the nature of man, universe, and God. At this
stage men move to a metaethical level to look at their moral
assumptions, rules, and principles to infer them from a broader picture
of the world as a whole. Moral principles are placed in a wider
context of metaphysical and religious views. For Kohlberg, the idea
of justice constitutes an essence of morality. However, at this stage,
men, says Kohlberg, look at this idea which was a guiding principle in
their life and substantiate it by seeing in it a consequence of the
constitution of the natural and (if admitted to exist) of the supernatural.
Unlike in other stages, when a distinction between the self and the
other, between *ego* and the world was constantly perceived and present
in man's reasoning, at stage 7 a holistic, or synthetic vision develops,
according to which "the self is understood as a component of this
order, and its meaning is understood as being contingent upon
participation in this order" (Kohlberg 1984, 250).

    The fact of introducing soft stages shows that Kohlberg is not very
restrictive about the use of cognitive-developmental method and sees
possibilities of extending and amplifying it. However, he may have a
tendency, like Piaget, to see the psychological world though cognitive
glasses, which may be somewhat detrimental to other areas of human
psyche. The area meant here is the affective and empathic side of man.
To be sure, Kohlberg includes affective elements in his discussion on
morality. His moral stages include references to fear, guilt, pleasure
and the like. However, they are not in the center of the scene,
because, according to him, it is not a proper place for them: "while
motives and affects are involved in moral development, the development of these motives and affects is largely mediated by changes in
thought patterns" (Kohlberg 1984, 63). In practice, the fact that affects
are "largely mediated" by cognition turns into total mediation, therefore

it is enough to concentrate on the latter with exclusion of the former. Kohlberg, after Piaget, mentions parallel development of affection and cognition (Kohlberg 1984, 9), but he considers cognition to be more perspicuous and exercising more influence in moral matters and hence, worthy of more attention than affection. This view may be considered a very strong argument supporting the view that the moral dimension is the main dimension of man which makes use of other dimensions to promote its claims and to expend the area of its influence. On the other hand, moral dimension may be viewed as restricted to the cognitive side, as one of areas of cognitive processes. Interestingly, Kohlberg seems to believe that in the course of moral development cognitive processes dislodge affections, and hence the higher is the level of moral development, the lesser place is occupied by affection in making decisions of moral nature. For instance, stage 6 subjects disobey authorities "in the name of individual rights, but without concern for individual rights involving empathy for a concrete victim" (Kohlberg 1984, 70-1). It may be claimed, then, that an ideal stage 6 (and 7) subject is a cold, calculating contrivance which infers some consequences from moral assumptions stored in its memory by its designer. Hence, it seems that treating the affective in man as a mere side-issue somehow regulated by the cognitive is a real weakness of an otherwise impressive theory of Kohlberg's.

But in light of such a treatment of affection it would be difficult to say why people act morally. It is not enough to know what are moral rules of behavior and what morality prescribes for a certain situation. There has to be the will, drive, or desire to act morally, desire motivated by the moral dimension, by affection and empathy, by making goodness the primary motive of actions (cf. also Hoffman 1970, 280). Also, to decide which course of actions to choose, each of them based on a different principle, cognition may be insufficient and morality has to be called for resolution. We can agree with Piaget that affectivity is an "energy source" which puts cognition into motion and causes that cognition's precepts are actually applied. It is having the good of others on mind that leads people to go even through hardships to bring about this good on individual and social levels. If this affective component is missing then it is seriously questionable whether some behavior can adequately be called moral, and rules standing behind it can be considered moral rules, even if they are dressed in moral terms. Is robot a moral entity? It 'knows' rules, since they are stored in binary code in its memory, but can we - without violating meaning of language - call its behavior moral, even if it leads to improvement of human life? Is moral a syringe with which a nurse injects a curative potion of medicine? Therefore, we may ask whether people who attained stage 6 and lost their empathy and sympathetic attitude really made a moral progress. Our contention is that cognition without affection and empathy cannot be considered morality. Moral behavior has to be affectively motivated, brought about by empathy to be called moral. Affection, this "energy source" fuels cognition to cause effective action, but cognition is a subsidiary of empathy, not vice versa. Some even claim that moral emotions, and not reasoning are the only element of morality that is common to all people; e.g., Kagan lists five core moral emotions: fear of punishment, empathy, guilt, *ennui* from the oversatiation of a desire, and anxiety caused by awareness of inconsistency between actions and beliefs. Such feelings would lead to acquisition of moral virtues characterizing a given historical situation (Damon 1988, 14).

Empathy (*Einfühlung*) is seen in different light by different authors, the central issue being its affective and cognitive nature. It is agreed that both elements participate in empathy, with the discussion centering around emphasis and priority assigned to each, both synchronically and diachronically. Is empathy originally affective, or cognitive? In adults, does the cognitive or affective element prevail? The cognitive-developmental tradition of Piaget's accentuates the cognitive component, so that there is a tendency to study empathy in older children, adolescents and adults whose cognitive functions are well developed. Signs of empathy in babies are ascribed to instincts or to some pre-empathic reactions (emotional resonance or contagion) which only outwardly resembles empathy. This approach makes a child, by birth, an egoist unable to 'feel into' someone else's inward states, since cognitive dimension is underdeveloped, and intuitive along with operational stages of cognitive development are apparently not sufficient for a child to develop genuine empathy.

Another approach to empathy is to bring affective and emotional dimension to the foreground and to see in this dimension a primeval force of cognitive development. In this way, affections truly become an "energy source" not only fueling development of cognition but also triggering it, enabling its emergence.

It has been observed by many that moral behavior can also be found in youngest children, in babies and toddlers, not only in adolescents and adults. Parents have already long ago observed that even two-day old babies behave as though they were hurt hearing other child cry. This reaction has been systematically scrutinized by Simner (1971), who played to newborn babies 6 minute tapes containing cries of other babies and a computer simulation of a baby cry. He found that babies reacted by crying significantly more frequently to cries of babies of their own age than to cries of older babies or to simulated cry. This study has been replicated with similar results (Thompson 1987, 125-6).

Older children expressed a similar reaction. A 9-month study of children 10-20 months old indicated that children react with cry to other children's cry. Also, children made an attempt to pacify distressed children; the older children were, the more helpful and more constructive their actions became. But good intentions were always present. For instance, 18-months-old boy leads his mother to a crying friend to comfort him, even though this friend's mother was there (Zahn-Waxler, Radke-Yarrow 1982; Thompson 1987, 131; Hoffman 1984, 285). This is an illustration of an empathic response caused by affection to other child's uneasiness, by feeling uncomfortable and the desire to help. It would indicate, as many scholars believe, that the potential moral emotions are inborn.

These moral emotions constitute a spark of humanity in newborns, and they are expressed through empathy from the earliest days of their lives, regardless of how unsophisticated the mechanism of empathy can be at that time. Empathy by itself, as observed by Heinz Kohut, is neutral and it can be used as much by perpetrators to achieve their destructive goals by empathically assessing victim's situation as by benevolent actors wishing to improve other's well-being (1980, 483). Therefore, an existence of empathy does not ascertain by itself a moral character of child. This moral character, or some of its elements, has to be already in a child before development even starts. However, its existence is not a guarantee that the moral child will become a moral adult, a statement which hardly can be challenged. There are many factors influencing development of sympathetic empathy and of moral dimension of man, and only some of them will be mentioned (see Barnett 1987 for fuller discussion and an extensive bibliography).

An important factor in moral development is an affective attachment between parents and children. Children characterized as securely attached at 15 months of age were more sympathetic to their peers' problems two years later than children with anxious and insecure attachment. Also, undergraduate students with affective parents were more empathic than students, whose parents were less affectionate during their childhood. Such parental affection positively reinforces development of empathy by satisfying children's needs.

However, it is interesting to observe that the amount of warmth and love does not seem to increase the level of moral development. Grinder (1962) conducted an experiment in which he measured children's resistance to temptation in the situation which required violating rules in order to win a game and receive a prize. He stated that "warmth as a background factor out of which dependency can develop has little significance upon resistance to temptation of pre-adolescents" (p. 812). However, all children participating in this research were from "intact families" (p. 817), therefore, it can be assumed that all (or most) of them were reared in an atmosphere of love and warmth, which ascertained that a certain level of moral development was already reached, but more love simply did not translate into acceleration in this development and, consequently, in better resistance to temptation (cf. Srampickal 1976, 140). "Extremely high warmth and complete absence of punishment do not seem to be particularly facilitating of moral development" (Kohlberg 1984, 75-6), but their absence or insufficiency will cripple this development by forcing child to develop slowly or even to regress.

Affection in parents not only induces affection in children through an emotional channel, but children also learn from parents using them as models for their behavior. Parents who sympathetically respond to children's concerns show children a proper reaction to other's distress. To be sure, children use not only parents as their models, TV being one of predominant factors in teaching sympathetic response, or no response at all. Having an adult model is of extreme importance since it slows down or accelerates the moral development. It is even said that Piaget's "developmental stages were readily altered by the pro-vision of adult models who consistently adopted moral orientations that ran counter to those displayed by the child. Increased consistency in the children's moral orientation resulted from consistency on the part of the model" (Bandura, Walters 1963, 128). That is, as already remarked by Kohlberg, regression in moral development is possible if circumstances systematically divert normal developmental process.

Another factor of moral development is a highly competitive environment. Highly competitive boys are less generous and less empathic than non-competitive boys, since most often competing is ego-oriented, designed to show that I am better, I can do it better, I am number one. This atmosphere is not conducive to sharing spirit, since sharing undermines competitiveness and increases a possibility of losing.

However, a positive self-image is needed for attaining an adequate level of moral behavior, since preoccupation with our own faults, real or imaginary, may lead to withdrawal and unwillingness to make contacts with others. This negative self-image can be broken by assigning children some responsible tasks, finishing which not only gives a sense of accomplishment, but also strengthens children's sensitivity. It is known to all parents that if children are caretakers of their younger siblings, they mature faster and become more responsible than children who do not perform such duties.

One of the most important factors in moral development is parental discipline. Hoffman indicates that the most efficient way for promoting internalization of ethical rules and in developing empathy in child was induction. This technique consists in pointing by parents, especially by empathy, to consequences of child's acts and to distress caused in others. Studies show that children whose parents used this technique were more generous and considerate of others than children of parents who used physical punishment. Induction is largely based on reasoning, since parents have to explain why certain behavior in unacceptable and causes harm in others. Explanation has to be clear and understandable. However, such a message has to be supported by display of emotion to be effective. "Merely encouraging an awareness of the affective state of another person through the use of a calm and well-reasoned explanation was found to be insufficient to elicit the child's emotions and attempts to help" (Barnett 1987, 153-4). Display of emotions indicates that we do mean what we reason about and that cold logic may have little impelling power by itself.

It is important to notice that induction was most efficient in children, because it is based on a child's need for love rather than on cognitive skills, and "the threat of love withdrawal implicit in inductions is relatively mild" (Hoffman, Saltzstein 1967, 235). When using this method, affection can still be retained and expressed, whereas in love-withdrawal, which consists in expressing anger or disapproval of some behavior, affections are at least not displayed whereby child is emotionally more threatened than in the use of reasoning whose effectiveness can even be reinforced by some display of affection. "Despite the diversity of theoretical approaches, measuring instruments, and moral content areas ... results have a common core of agreement: ... an internalized moral orientation is fostered by an affectionate relationship between the parent and child, in combination with the use of other techniques;" therefore, "psychological discipline which capitalizes on the affectionate relationship ... fosters the development of internalized moral structures in general" (Hoffman 1963, 305, 312).

All these facts indicate that moral development does not necessarily lead to moral personality and to forming altruistically inclined persons. Moral development is sensitive to a variety of influences and moral dimension can properly develop only in appropriate circumstances. If circumstances are not favorable, the moral dimension can be suppressed and, as an outcome, a delinquent personality can be formed.

Delinquents come usually from homes where severe corporal punishment was exercised, often done in anger and unproportional to the misdeed. Research shows that aggressive behavior of children is significantly correlated with physical punishment. Children, who were frequently threatened with consequences of their misbehavior were more likely to became aggressive. 95% of aggressive boys were rejected (treated cruelly, openly disliked) by at least one parent (McCord, McCord, Howard 1961).

Deprived neighborhoods, little parental supervision, association with deviant peer groups from early childhood, and broken or unstable homes are all factors of delinquent character. Little parental supervision oftentimes means parent's disinterest for children's life. In the short run, children may even appreciate it, but in the long run it has catastrophic consequences. Broken homes do not have to lead to delinquent behavior, if atmosphere at home is good. Research shows that unbroken but unhappy home results in a higher rate of delinquency than a broken but happy home (Nye, Short, Olson 1958). Delinquents, according to other studies, receive less affection from their parents than non-delinquents and are more inclined to feel that parents are unconcerned about their well-being (Glueck, Glueck 1950). As a consequence, rebellion associated with serious misconduct often results.

Development of moral dimension "requires the use of authority, firmness, planned limitation, and at times punishment," however, "what is necessary for success is, first of all, a warm accepting attitude on the part of the parent or parent substitute" (Hewitt, Jenkins 1946, 86). But because this element is missing or is very weak between the parent and later-to-be-delinquent, the development of morality is thwarted at the outset. No amount of rational argumentation and reasoning on the part of teachers or counselors would suffice to reverse the process of conscience malformation and to retract maleficent results of parental mistreatment, be it neglect or rejection. The atmosphere of warmth and love at home is a necessary condition for a proper development of other-orientation. The need of love is basic to children and if love is withdrawn from them, an irreparable damage is done.

Love for children must not be equated with a permissive atmosphere, since such an atmosphere can have very undesirable effects on moral development. Lack of discipline on the part of parents may be very close to disinterest with regard to children and their deeds, and will be interpreted as lack of love. Hence, not surprisingly, love-oriented disciplinary methods were positively correlated to guilt feelings in 5-6 years old children. This discipline should not be erratic, since inconsistency may lead to development of aggression in children. Also, placing high demands on children is correlated with non-aggressiveness of children, whereas parents making low demands have aggressive children (McCord, McCord, Howard 1961).

Parents have to find a proper balance between discipline and permissiveness on one hand, and between treating children as equals and treating them in an authoritarian way on the other. Alfred Baldwin "found that children from families with democracy without control were often cruel and disobedient, while children in families with control without democracy often lacked initiative and an inner sense of responsibility. The combination of democracy and control in the family led to an optimal pattern of assertive kindliness on the part of the child" (Damon 1988, 57).

Before this chapter is brought to a close, one very important assumption has to be explicitly stated. With Rousseau, Fichte, and Carl Rogers (1961, 91) I believe that man is good by nature and this goodness is embodied in inborn moral emotions and in an ability to love, which are manifested through empathy when a proper situation arises. This goodness of man's nature causes that goodness is the main goal of man and moral dimension constitutes the core of humanness.

Only good can engender good, therefore, a natural tendency is to positively develop a moral dimension using other dimensions as subsidiaries, in particular, cognition. Cognition plays a crucial part in development of moral dimension without which the latter would be largely ineffective. By nature child is not an egoist, but this non-egoistic nature is expressed by infants in ways which do not alleviate someone else's problem or may be even harmful to others if cognition is not included. Children have good intentions with extremely limited knowledge of how to express them in a helpful way. Therefore, cognition has to be sufficiently well developed in order to implement effectively these intentions. From the beginning, the process of cognitive development is fueled by an affective dimension in a child, an "energy source" which not only makes this development possible, but also endows it with meaning. Cognitive development for its own sake is destructive in the long run, and should only be a means and not the primary goal. Therefore, affections are not only an energy source, but also a value injector, which causes moral development analyzed from cognitive-developmental perspective to move upward from self-orientation to other-orientation, from self-centeredness to social perspective. This upward movement occurs, let us repeat, if seen from cognitive-developmental perspective. On the other hand, from an affection-oriented perspective, infants have from the very outset very strong altruistic tendencies which erupt in affective outbursts and only because of a child's intellectual deficiencies, these tendencies may be assigned by researchers an insufficient weight or altogether disregarded. A child's affections are much more developed than the cognitive dimension, and these affections, so to say, have to wait until cognition reaches sufficiently developed stage to be used by the moral dimension effectively. When confined to the cognitive-developmental perspective, moral dimension may seem lagging behind cognitive development, since analysis of morality in child is limited to moral reasoning, and, of course, moral reasoning can take place if reasoning as such exists. Therefore, moral *reasoning* depends on cognitive development. However, even Piaget, who sees children as egocentric and self-centered creatures admitted that spontaneous affection "prompts the child to acts of generosity, and even of self-sacrifice, to very touching demonstrations which are no way prescribed" (1929, 195). Is it what we should expect an obedience-driven and egocentric child to do? Therefore, I would like to take an issue with the statement that there exist "certain fundamental and 'natural' cognitive bases of moral emotion" (Kohlberg 1984, 67), since the dependence seems to be in opposite direction: moral emotion is a basis of cognition which emerges in order to support and maintain these moral emotions. Moral emotions without cognition are inept, cognition without moral emotions is void.

The cognitive-developmental perspective proved more than any other perspective that there is a universal moral development which gradually opens men to others by integrating them with the world outside the self. However, by concentrating on cognitive dimension, this perspective did much injustice to the affective dimension seeing in it less significant dimension of man. But, as Piaget admitted, without affections cognitive development would not take place, since they are an energy source for this development. This statement can be strengthened if in the spirit of Hume we say that reason is a slave of emotions. Affections enable cognitive development to be its tool used to externalize in a serviceable and helpful way goodness in man. There is a time lag between moral and cognitive development, but it is cognitive development which is behind, not vice versa. It is cognitive perspective which makes man egocentric and only after attaining certain level man is a fully moral being on both affective and cognitive levels. However, it is a years-long process during which child is exposed to variety of influences which may affect this process, both positively and negatively. Goodness in children is not cast in iron; it is a fragile construct which has to be cherished by the self as much as by others. And because a child is a helpless and receptive being, at the beginning moral development is especially sensitive to outside influences, which may strengthen it, accelerate, or slow it down and even suppress. A proper atmosphere is indispensable, intellectual, but first of all affectionate, since, as observed, moral development, even if analyzed from cognitive perspective, stagnates when limited to cognitive dimension, because reasoning is almost never effective, unless supported by other methods. Affection on part of parents or caretakers has to be present to sustain in children the affective dimension and promote an upward movement of cognitive dimension, affection which is expressed in love.

Love is essential for human and humane life and for promoting moral (and cognitive) development. As Dieter Eicke indicates it, according to most psychologists the primary drive in humans is "the need to be loved, to enjoy approval and affection" (Glaser 1971, 170), the need to be loved, not the need to be reasoned to. Love induces love, parental love spurs the capacity of love in children to bud and to flourish. The capacity of love remains dormant if not awaken by love of others, in particular parents. This also is a foundation of ethical dimension, since - as argued by Erich Fromm - affective maternal conscience is based on child's own capacity to love (Fromm 1956, 43-44). If love fails, everything else fails as a consequence, since love is "the ultimate and real need in every human being" (p. 133). Even Freud admitted at the end of *The ego and the id* that "living means the same as being loved. "

Need of being loved is a primary human need, and if fulfilled, it enables man to love others, which implies acting for the good of others, bringing happiness and well-being to others, especially to the loved ones, to act unselfishly and altruistically. Love is the primary affection which makes man a moral person, elevating the moral dimension in man to the highest level. Existence of the capacity to love rooted deeply in the self means that man is good by nature. The capacity of love, awaken and livened up by love, is a basic ingredient of human personality. Its development is development of personality. If this capacity is crippled, an unhealthy personality emerges, deviant, maybe even psychotic - unable to feel affection, unable to relate to other people (cf. description given by Stephenson 1966, part 2). If this capacity blossoms, man's moral dimension grows to the benefit of others - and to their joy.

References

Bandura Albert, McDonald F. 1963, The influence of social reinforcement and the behavior of models in shaping children's moral judgment, *Journal of Abnormal and Social Psychology* 67, 274-281.

Bandura Albert, Walters Richard H. 1963, The generality of moral behavior, in Johnson et al. 1972, 125-130.

Barnett Mark A. 1987, Empathy and related responses in children, in Eisenberg, Strayer 1987, 146-62.

Bergling Kurt 1981, *Moral development: The validity of Kohlberg's theory*, Stockholm: Almquist & Wiksell.

Boehm Leonore 1957, The development of independence: A comparative study, in Johnson et al. 1972, 201-7.

Burton Roger V. 1984, A paradox in theories and research in moral development, in Kurtines, Gewirtz 1984, 193-207.

Carter Robert 1985, Does Kohlberg avoid relativism?, in Modgil, Modgil 1985, 9-20.

Cohen David 1983, *Piaget: Critique and reassessment*, London: Croom Helm.

Damon William 1984, Self-understanding and moral development from childhood to adolescence, in Kurtines, Gewirtz 1984, 109-27.

Damon William 1988, *The moral child: Nurturing children's natural moral growth*, New York: The Free Press.

Damon William 1977, *The social world of the child*, San Francisco: Jossey-Bass.

Eisenberg Nancy, Strayer Janet (eds.) 1987, *Empathy and its development*, Cambridge: Cambridge University Press.

Flavell John H. 1963, *The developmental psychology of Jean Piaget*, New York: Van Nostrand.

Freud Sigmund 1930, *Civilization and its discontents*, New York: Norton 1961.

Freud Sigmund 1933, *New introductory lectures on psychoanalysis*, New York: Norton 1965.

Freud Sigmund 1940, *An outline of psychoanalysis*, New York: Norton 1969.

Fromm Erich 1956, *The art of loving*, New York: Harper & Row.

Getzels Jacob W., Jackson Philip W. 1962, *Creativity and intelligence: Explorations with gifted students*, London: Wiley.

Glaser John W. 1973, Conscience and superego: A key distinction, in Nelson C. Ellis (ed.), *Conscience: Theological and psychological perspectives*, New York: Newman Press, 167-88.

Glueck Sheldon, Glueck Eleanor 1950, *Unraveling juvenile delinquency*, New York: Commonwealth Fund.

Grinder Robert 1962, Parental childrearing practices, conscience, and resistance to temptation of six-grade children, *Child Development* 33, 803-20.

Handlon B.J., Gross P. 1959, The development of sharing behaviour in children, *Journal of Abnormal and Social Psychology* 58, 425-8.

Heath Douglas 1977, *Maturity and competence: A transcultural view*, New York: Gardner.

Hewitt Lester E., Jenkins Richard L. 1946, *Fundamental patterns of maladjustment: The dynamics of their origins*, Springfield: State of Illinois.

Hoffman Martin L. 1963, Child rearing practices and moral development: Generalizations from empirical research, *Child Development* 34, 295-318.

Hoffman Martin L. 1982, Development of prosocial motivation: Empathy and guilt, in Eisenberg N. (ed.) *The development of prosocial behavior*, New York: Academic Press.

Hoffman Martin L. 1984, Empathy, its limitations, and its role in a comprehensive moral theory, in Kurtines, Gewirtz 1984, 283-302.

Hoffman Martin L. 1970, Moral development, in Mussen Paul H. (ed.), *Carmichel's manual of child psychology*, vol. 2, New York: Wiley, 261-359.

Hoffman Martin L., Herbert D. Saltzstein 1967, Parent discipline and the child's moral development, in Johnson et als. 1972, 221-37.

Johnson Ronald C., Dokecki Paul R., Mowrer O. Hobart (eds.) 1972, *Conscience, contract, and social reality: Theory and research in behavioral science*, New York: Holt, Rinehart & Winston.

Kay A. William 1969, *Moral development: A psychological study of moral growth from childhood to adolescence*, New York: Schocken Books.

Kohlberg Lawrence 1971, From is to ought: How to commit the naturalistic fallacy and get away with it in the study of moral development, in Mischel T. (ed.), *Cognitive development and epistemology*, New York: Academic Press.

Kohlberg Lawrence 1964, Development of moral character and moral ideology, in Lefrancois Guy R. (ed.), *Little George*, Belmont: Wadsworth 1974, 446-83.

Kohlberg Lawrence 1980, Educating for a just society: An updated and revised statement, in Munsey Brenda (ed.), *Moral development, moral education, and Kohlberg*, Birmingham: Religious Education Press, 455-70.

Kohlberg Lawrence 1984, *The psychology of moral development: The nature and validity of moral stages*, San Francisco: Harper & Row.

Kohlberg Lawrence 1976, Moral stages and moralization: The cognitive-developmental approach, in Lickona T. (ed.), *Moral development and behavior: Theory, research, and moral issues*, New York: Holt, Rinehart and Winston.

Kohlberg Lawrence, Gilligan Carol 1971, The adolescent as a philosopher: The discovery of the self in a postcoventional world, *Daedalus* 100, 1051-86.

Kohut Heinz 1980, Reflections, in Goldberg A. (ed.), *Advances in self psychology*, New York: International Universities Press, 473-554.

Kurtines William M., Gewirtz Jacob L. (eds.) 1984, *Morality, moral behavior, and moral development*, New York: Wiley.

Liebert Robert M. 1984, What develops in moral development, in Kurtines, Gewirtz 1984, 177-92.

Locke Don 1985, A psychologist among the philosophers: Philosophical aspects of Kohlberg's theories, in Modgil, Modgil 1985, 21-38.

Macaulay E., Watkins S.H. 1925-6, An investigation into the development of the moral conceptions of children, *The Forum of Education* 4.

McCord W., McCord J., Howard A. 1961, Familial correlates of aggression in non-delinquent male children, *Journal of Abnormal and Social Psychology* 62, 79-93.

McDougall William 1918, *An introduction to social psychology*, Boston: Luce.

Modgil Sohan, Modgil Celia (eds.) 1985, *Lawrence Kohlberg: Consensus and controversy*, Philadelphia: The Falmer Press.

Mowrer O. Hubart 1967, Conscience and the unconscience, in Johnson et al. 1972, 349-71.

Nisan Mordecai 1984, Content and structure in moral judgment: An integrative view, in Kurtines, Gewirtz 1984, 208-224.

Nye F. Ivan, Short James, Olson Virgil J. 1958, Socioeconomic status and delinquent behavior, *American Journal of Sociology* 63, 381-9.

Osborne F.W. 1894, The ethical content of children's minds, *Educational Review* 8, 143-6.

Piaget Jean 1929, *The moral judgment of the child*, New York: Free Press 1965.

Piaget Jean 1954, *Intelligence and affectivity: Their relationship during child development*, Palo Alto: Annual Reviews 1981.

Piaget Jean 1971, Closing remarks, in Green D.R., M.P. Ford, G.B. Flamer (eds.), *Measurement and Piaget*, New York: McGraw-Hill, 210-213.

Piaget Jean, Inhelder Bärbel 1951, *The psychology of the child*, New York: Basic Books 1969.

Pittel Stephen M., Mendelsohn Gerald A. 1966, Measurement of moral values: A review and critique, in Johnson et al. 1972, 79-97.

Rogers Carl R. 1961, *On becoming a person: A therapist's view of psychotherapy*, Boston: Houghton Mifflin.

Sharp F.C. 1898, An objective study of some moral judgments, *American Journal of Psychology* 9, 198-234.

Simner N.L. 1971, Newborn's response to the cry of another infant, *Developmental Psychology* 5, 136-50.

Snarey John R. 1985, Cross-cultural universality of social-moral development: A critical review of Kohlbergian research, *Psychological Bulletin* 97, 202-32.

Srampickal Thomas 1976, *The concept of conscience*, Innsbruck: Resch.

Stephenson Geoffrey M. 1966, *The development of conscience*, London: Routledge.

Sugarman Susan 1987, *Piaget's construction of the child's reality*, Cambridge: Cambridge University Press.

Swensen Clifford H. 1962, Sexual behavior and psychopathology: A test of Mowrer's hypothesis, in Johnson et al. 1972, 372-6.

Thompson Ross A. 1987, Empathy and emotional understanding: The early development of empathy, in Eisenberg, Strayer 1987, 119-45.

Tomlinson-Keasy C., Keasy Charles B. 1974, The mediating role of cognitive development in moral judgment, *Child Development* 45, 291-8.

Turiel Elliot 1966, An experimental test of the sequentiality of developmental stages in the child's moral judgment, in Johnson et al. 1972, 308-18.

Ugurel-Semin R. 1952, Moral behaviour and moral judgment in children, *Journal of Abnormal and Social Psychology*, 463-74.

Weiss Daniel, Richard J. Haier, Daniel P. Keating 1974, Personality characteristics of mathematically precocious boys, in Julian C. Stanley, Daniel P. Keating, Lynn H. Fox (eds.), *Intellectual talent: Discovery, description and development*, Baltimore: Johns Hopkins University Press, 126-39.

Whyte Lancelot L. 1960, *The unconscious before Freud*, New York: Basic Books.

Zahn-Waxler Carolyn, Radke-Yarrow Marian 1982, The development of altruism: Alternative research strategies, in Eisenberg Nancy (ed.), *The development of prosocial behavior*, New York: Academic Press.

Chapter 5

Moral Dimension of Man
and
Education

Socrates reportedly said that education (*paideia*) was "the greatest good of men" (Xenophon, *Apology* 21). Also today the importance of education is invariably recognized and educational issues are used in political campaigns. If education is ascribed such an elevated status, both in Antiquity and the twentieth century, and, on the other hand, as claimed in the preceding chapters, the moral dimension defines man, then we should ask, what is the relation between the two? What role does the moral dimension play in education and what role should it play?

The moral dimension is implicitly presumed in stating the goals of education. If it is said that "the goal of education ... is ultimately the 'self-actualization' of a person, the becoming fully human, the development of the fullest height that the human species can stand up to" (Maslow 1971, 169, similarly Erich Fromm), then very important assumptions are made concerning human nature. In particular, would it really be a desirable goal to support self-actualization, if man were a bundle of instincts, among whom libido prevails? If libidal drive constituted the core of human nature, then the best thing education could achieve would be suppressing self-actualization. If man's nature were selfishness (Hobbes), or drive for power (Adler), then allowing man to become fully human would be self-destructive for humanity. That is why education and acculturation was viewed as suppressive and oppressive (Hobbes, Freud), or simply as training and taming. If the goal of education is to help man to become fully human, and if its aim is man's self-actualization allowing full development of human potential, then the underlying assumption has to be that man is good by nature. If this nature were evil, destructive, selfish, egotistic, etc., then self-actualization would be accessible to the privileged few at the cost of others' non-actualization. If possessing virtue "means to be fully and adequately what one is capable of becoming" (Dewey 1922, 415), then it ought to be assumed that goodness lies at the heart of human nature. This goodness, which is worth developing, expanding, and actualizing to enable it to become a leading force in human life.

If goodness constitutes human nature and education aims at self-actualization, then developing this goodness becomes the center of education. This goodness is understood in terms of altruism, other-centeredness, caring, and hence the "primary aim of every educational institution and of every education effort must be maintenance and enhancement of caring" (Jarrett 1991, 66). Thus, cognitive and cultural aims are not primary. Although intellectual rigor is indispensable, it is caring which is central in moral education. Knowledge becomes alive when caring is associated with what is known.

Today this cognitive and cultural aim comes to the forefront and moral dimension, if not treated instrumentally, can become an aspect which should be analyzed exclusively in rational terms similarly to man's physiology. This can be observed in the cognitive-developmental approach which induced the reasoning approach to moral education. Dilemmas are presented to the students who discuss them and provide reasons for a specific behavior, and justification for a certain course of action. Moral dilemmas are resolved by reasoning, morally difficult problems are solved using almost a deductive mode of thinking, and a moral discussion is characterized by "an intense focus on the moral reasoning aspects of a moral conflict" (Hersh et al. 1979, 129). Stressing rational aspects of moral problems may lead to dissolution of morality in analyses of presuppositions and consequences from the standpoint of logical resolution. Substantiation in terms of the affective and empathic reasons may sound almost irrational, and hence of little weight. A recourse to the subjective is hardly acceptable, because of what it is - the subjective, that is unreachable for intersubjective clarification and verification, it is simply unscientific. What is moral and personal becomes irrelevant, and hence, exercising little or no influence on human behavior. The rational domain gains dominion by suppressing the non-rational. The moral arguments play no role since that seems to be irreconcilable with the spirit of time which requires rationality to have the last word.

The core of each person is what Augusto Blasi called after Kant the good will: "the central affective and motivational orientation to morality that characterizes each person," an orientation "towards virtue, justice, altruism, briefly, towards the moral as each person understands it" (Blasi 1985, 435-6). However, there is a tendency to move away from this understanding of personality and its elements. It is achieved by "an overwhelming preference for morally neutral instrumental variables as objects of study," so that the good will is replaced by ego strength, by naturalization, so that the moral characteristics are said to originate from biological make-up or from experience (p. 440).

Already a century ago John Dewey has remarked that "every teacher requires a sound knowledge of ethical and psychological principles ... Only psychology and ethics can take education out of the rule-of-thumb stage and elevate the school to a vital, effective institution in the greatest of all constructions - the building of a free and powerful character" (Dewey, McLellan 1895, 207). This observation is equally true in kindergarten and in graduate school. At no level is education merely about acquiring new theoretical and technical information, or about knowledge - it is about wisdom: about knowing how to live with others, how to become a part of society, how to implant own life with meaning, and how to be useful to others. Education is about unlocking dormant goodness in man, cherishing it, and supporting until it can stand on its own feet, to be in service to others. Education is developing humanness in man, rearing the other-orientation, blowing up the flickering spark of the good core which constitutes man. If a teacher, at any level of education, forgets it, the school becomes a cold and spirit-less institution for transferring chunks of information from one information storage to another, from one brain to another. If stuffing students with knowledge becomes the primary goal of school, then the purpose of education is defeated and training of subjects takes the place of educating students.

To be sure, this goal can be practically embodied in a variety of forms and it should be treated differently at different levels of education. Whereas a warm, accepting, and lovable attitude is very important at each level, it is of crucial importance at the first stages. A cold and insensitive teacher in kindergarten and primary school can make irreparable damage. On the other hand, love and acceptance can boost development, both intellectual and social, in children who otherwise might be treated as a lost cause. Sidney B. Simon and Robert D. O'Rourke describe their approach to "troubled, difficult, failure-oriented children" at Boxelder School; the crux of this approach was to cause all children to "feel important, cared for, listened to, and loved as people who are learning to extend themselves in loving ways ... These good things happen with children when teachers care as much about nurturing healthy self-respect in their students as they do about educating them" (1977, 9).

At later stages of education the emphasis on developing wisdom and goodness in man is not as strong as during the initial stages, but it never disappears, or at least it should not. Education should continue with understanding that the knowledge gained in school is not a goal in itself and that maintaining wisdom is the purpose of acquiring knowledge. Even the best of intentions can be harmful if carried into effect with insufficient knowledge. Therefore, the secondary school and university stress the knowledge aspect of education more than previous stages. However, the ultimate goal must never be lost sight of, otherwise education becomes a vain and meaningless exercise in memorization. Unfortunately, many a time such a tendency may be observed, and this is eventually damaging for each individual and for the society as a whole. Education is not about extending horizons of knowledge, it is about making the moral dimension of man effective. The rational dimension should invariably be subordinate to the moral dimension, the latter should be an assumption, the ever present goal - possibly unspoken and rarely mentioned - of education at any level. Therefore, the university should be the last institution which proclaims that "science should not be hampered by ethical judgments" and which drives a wedge between cognitive and moral dimension by stating that some concerns, such as the use of animals in research, "is not an intellectual concern - it's an emotional, an ethical one, and a moral one" (quoted by Rollin 1991, 152). This tendency to separate scientific concerns from ethical issues seems to be very strong in the university, which eventually is very detrimental from the educational standpoint. The message to the students is that when some scientific research is technologically feasible then this very fact suffices as a substantiation for proceeding with this research. Other concerns are less important, even harmful, or at most irrelevant, therefore what science proclaims as possible, should become real. It is known that science strives for truth, therefore obtaining truth should be at the top of agenda of every academic endeavor, regardless of other considerations.

Ethical questions are frequently suppressed or removed from the scene when it comes to scientific work. There is a division of labor between scientists and moralists, and the separating wall between them is becoming higher and thicker, with very disastrous consequences. A good example is the Manhattan Project for developing an atomic bomb, which continued, even after it became clear that the Germans were unable to produce a bomb. Although some scientists wanted to halt it and some left it (eg. Joseph Rotblat), the Project acquired its own momentum. The scientists were interested in the project itself, guided by "pure and simple scientific curiosity," and most of the scientists "were not bothered by moral scruples; they were quite content to leave it to others to decide how their work should be used" (Rotblat 1985, 18). As a result of such an attitude, we can find a pronouncement that "science has nothing to be ashamed of even in the ruins of Nagasaki" (Bronowski 1955, 73). Can science pretend innocence and claim it was just exploring the truth? If so, it is the innocence of a child who does not realize that some actions can have harmful consequences. "The scientific spirit is more human than the machinery of governments," the latter being blamed for Nagasaki, but if science willingly subjugates itself to the rules of this machinery, then the scientific spirit loses its human dimension. This happens if science loses sight of the good or well-being of man, and the search for truth becomes its only motive, whereby the problem of violating moral principles by scientific research becomes imminent.

This attitude is not an isolated happenstance. A scientist participating in developing software for SDI poses in passing some moral questions, but immediately dismisses them by stating: "I am not an expert on moral and political issues and offer no answers to these questions" (Parnas 1985, 609), and many a scientist would endorse this rationale. Ethics is good for ethicists, maybe even for priests, but not for scientists, who quasi-humbly confess that they are not experts. But who is an expert in moral issues after all? If a renowned scientist ducks such problems, what is to be expected from those whose education does not match his?

Scientists want to concentrate on their projects and forget the context in which the projects are or will be used. They do not want to see some potential problems between this context and the results of the projects. This separation line leads to a disdain with which including an ethical discussion with regard to some project is considered.

Why is it, however, that scientific and technological problems are analyzed from an ethical perspective only infrequently? The reason lies, it seems, in seeing science and technology as an engine of progress. In fact, there are two philosophies in this regard: optimistic and pessimistic.

In an optimistic outlook, science and technology is the primary, if not the only, cure of all social ills, the only way of creating a humane future, the only hope of man. Technology, by creating more contrivances, faster devices, and more efficient artifices, rises hope that there will be less hunger in the world, more energy, that people will suffer less, there will be more happiness, and that human relations on the family, country, and international levels will be more harmonious.

This view is a fruit brought by the Industrial Revolution and was very common at the turn of the century. "We should not be disturbed in our faith - wrote once Werner Siemens - that our zeal in investigation and discovery will raise mankind to higher grades of civilization, will ennoble it and make it more amenable to ideal efforts, and that the dawning scientific age will diminish its suffering and disease, heighten its enjoyment, and make it better, happier, and more satisfied with its lot" (Siemens 1886, 23-4). In that period there was a vision of a "scientific way of life," and even novels were written according to a scientific method (Émile Zola). This high regard for science and technology lasted until 1945, when Hiroshima showed how destructive achievements of science could be. However, with the spreading of computers, science and technology regained their high status, especially among technocrats. However, they are also venerated by humanists, as expressed, for instance, by theologian Myron B. Bloy, who almost repeats the statement of Siemens: "technology is a stage in man's development towards ... maturity. Technology is first of all a means of Western man's growth into a deeper measure of freedom, for it has freed us from an absolute preoccupation with survival and has opened up a whole range of possibilities, such as justice, friendship, political an economic equality, and esthetic enjoyment" (Bloy 1966, 332). Thanks to science and technology, humanity moves upward to a better and happier world. If it were not for science and technology, cruelty would prevail, unhappiness would be rampant, enjoyment impossible, friendship rare, and equality a utopia.

Such promulgations are by no means isolated and are pronounced by many scientists. For instance, Robert Jastrow writes: "the computer will minister to our social and economic needs. Child of man's brain rather than his loins, it will become his salvation in a world of crushing complexity" (quoted in Hanson 1982, 309; cf. also Jastrow 1981, 162). This salvation may be in the form of peculiar incarnation: "At some point - says Marvin Minsky - people may even prefer to convert themselves into machines, because if they can transfer their intelligence into another embodiment, they might be able to live forever and continue developing" (quoted in Hanson 1982, 310; cf. also Jastrow 1981, 166-7). For those who prefer to stay in their bodies, a consolation can be found in I.J. Good's forecast that advances in medicine achieved through ultra-intelligent machines will allow even those alive today to live a thousand years (Evans 1979, 265). Small wonder that it may happen that "computers will be seen as deities, and ... there may even be an element of truth in the belief" (Evans 1979, 262).

For those who are not interested much in eschatology and soteriology, technology has much to offer as well. In the words of Christopher Evans, "the production of fantastically cheap devices ... will, at long last, make the humanistic dream of universal affluence and freedom from drudgery a reality". Also, "the increased affluence of the computerized world will inevitably spill over into the less-developed quarters." Humanity as a whole will be benefitted by the fact that machines and computers will do the slave work (Evans 1979, 243, 252, 258).

Not surprisingly, in many pronouncements of this kind there is a mention of the computer. The computer is merely fifty years old and the development of computer technology is simply remarkable. No other technology ever progressed at that pace, and no other technology became so ubiquitous. Therefore, the future of man is seen under the shadow of the computer, and, as the quotations above exemplify it, in bright colors. All possible downsides of technological development are left out from the picture, downplayed, or at least made minuscule in comparison with all the gains. Such proclamations are an expression of an over-optimism (or at least they easily slip into it) and the belief that all good will be brought to this earth automatically by the nature of technological development and will be eventuated through the natural course of scientific progress. No additional input is necessary; just caring for the scientific progress will suffice. As a result, ethical discussions are not needed, since everything develops toward the better and peaceful world anyhow. The morals will be improved, the humanity dignified, civilization made humane. Therefore, is there any need to discuss moral implications of scientific research? Not any more.

At the other extreme, there is a pessimistic view of technology verbalized, for instance, by French scholar Jacques Ellul.

Ellul assigns to technology a status of the fundamental force shaping the present and the future of humanity. Technology creates a basis upon which all other areas of life rests. It is a foundation without which the whole of the social life crumbles. Technology is a determining force driven by its own internal laws, it is impregnable to any influences except its inner impetus. Technology determines everything and nothing can affect its course. Therefore, trying to influence technology by political, economic, or any other means is irrational. Also, relating ethical issues to technology is out of place. At one point Ellul makes a sobering statement that "all the studies of medical and scientific ethics are useless and absurd". He says that the central object of all culture is the question of meaning and values, including moral values, whereas technology "rejects any relation to values. It cannot accept any value judgment ... about its activities," since its criteria are of a different nature. Therefore, "there can be no bridge between them. To associate them ... is nonsense" (Ellul 1990, 148).

Ellul is a total pessimist with regard to a possibility of exercising any influence on technology. However, he agrees at one important point with optimists: he promotes a thesis that techniques are the driving force of social development, as the quoted optimists, although they do that only implicitly and in a much less sophisticated way in a manner that fits Ellul's description of technocrats as "touchingly simplistic and annoyingly ignorant." However, the thesis itself is not that novel.

According to historical materialism, production forces determine everything else: the social, ethical and political systems, religious beliefs, laws, etc. The development of production forces is independent of the will of man and is governed solely by their internal laws. These laws determine the future of the society, and man has to adopt to them; otherwise, he becomes an obscurantist, or even an obstructionist. These laws inevitably determine the direction of social development, however, the latter can be sped up, or adjusted, if properly recognized and assessed. And the best judge of it is a communist party, the sole repository of wisdom concerning the trends of developments in history of production forces, production relations, and ideological superstructure.

Ellul's thesis on the determination of social development by techniques is comparable to the thesis on determination of production relations by production forces, although it is not, by any means, identical to the latter. What is identical is singling out a certain domain of human activity that acquires an independent status to the extent that its development cannot be controlled by man in any way. It becomes independent, not subject to the human steering power, and becomes alienated from man. Any effort to bring it under man's submission is illusory, since it is man who is controlled by this domain. Man can either march with the course of history or suffer from maladjustment.

The social organism is an extremely complex and delicate tissue. There are innumerable interdependencies between various elements and parts of this whole; more or less obvious, direct and indirect. These parts can be analyzed in their environment, or, if needed, independently. Thus, for instance, if one is interested in the development of railroad trains, airplanes, cars, or, generally, in different means of transportation, then a reference to moral values and the problem of good and evil may sound rather farfetched. However, in the long run, the statement that to associate the development of science and technology with culture is nonsense - is nonsense. Technology never develops in a social vacuum; it is connected by a multitude of links with political, economic, and ideological subsystems of the social whole. Ellul says that these links are unidirectional, from technology to everything else, thereby determining all that is beyond its sphere. Technological progress, says he, is equated with civilizational progress, and with the progress proper. Progress is to consist in producing more, faster, and more efficiently. Efficiency is the word of the day, a yardstick that measures progress, and technology is the means of increasing it. Therefore, what is in the way of technology, whatever hampers the increase of efficiency is, sooner or later, crushed or shoved aside, declared obsolete and ridiculous.

An example of such an obstacle is morality as defined by tradition or a religious system. The result is "the suppression of moral judgment, with the creation of a new ideology of science" (Ellul 1990, 19). The role of religion, to be sure, dramatically changed over the last hundred years. However, this change concerns mainly its institutional part; moral values are still alive and well, and they are transmitted now not mainly through the means of the religious education, but also through secular education, arts, and the law. The value of human life is still one of the most cherished values of man (at least in the Western civilization), as well as the esteem for parents, respect for someone else's property, etc. And these are values that can have a serious impact upon the development of technology.

One example of this impact is a failure of introducing for the first time firearms in Japan. In 1543 they were brought to Japan by some Portuguese adventurers and were used for some hundred years. Despite an obvious advantage of using muskets over swords and bows, they did not win a popularity, since they came into conflict with traditional values of honor, samurai pride, and the dignity of the fight. They killed too efficiently, that is, too soon, so that anybody could use them and no special skills were needed to shoot and kill. Abandoned, fire-arms were introduced for the second time after opening Japan to the West in 1853 (Boorstin 1991, 61).

Another example is the model technological efficiency exercised in concentration camps during WWII. We can just stand in awe seeing that nothing has been wasted from people "processed" in these camps. There were separate heaps of hair and glasses, human skin was used to produce some ware, human body to produce soap, golden teeth were removed and salvaged, and human specimens were used for numerous medical experiments, such as testing peoples' resistance to heat, cold, pressure, or various poisonous substances. However, despite this technological progress, the world condemned this way of treating people and this way of conducting experiments. The Nuremberg trials showed that in the eyes of the majority of men certain ways of techno-logical development are simply prohibited and closed - in the name of higher values and in the name of morality and human decency.

Therefore, the technology does not have to be a sole driving force of the social life, and, after all, Ellul agrees with that. "What will finally make it inevitable is neither the development of science and technique nor economic needs but the shaping of people who can do nothing else and will not be comfortable in any other society" (Ellul 1990, 400). What Ellul here means is an education system geared more and more to adjust students to the technological society more than to forming a person for whom producing more and faster does not constitute the meaning of life. However, it should be an education saturated richly with humanistic and moral values, and with the perception that the others are as important as one's own well-fare. It should form an attitude of posing questions concerning a sense of parti-cular developments, projects, and goals, and instilling a position of rejecting what impairs the well-being of man. Ellul sees such a chance but he chooses not to believe in a possibility of bringing it into motion.

There is a chance, though, and it is a good one. But an over-pessi-mist will not even try. Why try if all attempts are doomed to failure? And because *les extrêmes se touchent*, an over-optimist will not try to do anything either, believing that everything will be a happy consequence of the progress.   For both there is waiting: for the pessimist - painful and fearful, for the optimists - joyous and full of expectations.   But the sacrosanct technology will bring the future by itself and our attempt of influencing it can only taint it.

But the future is a *carte blanche* that will be filled up in accordance with our present actions.   The future is not foreordained.   We are not predestined to doom and gloom, or to joy and happiness; the future has to be forged by ourselves.   But technology and science by themselves will not suffice.   At all stages of their development there has to be a concern regarding goals, meaning, possible directions, and impact onto man.   That is why stressing moral issues at all times is of such importance.   The more we are surrounded by technology, the more we need non-technological dimension, the more the need for values going beyond producing more and faster.   It is a responsibility of the university to add as much human touch as possible to the educational process, and the university, by using its prestige, should promote, in theory and practice, the view that the goal of education is developing moral dimension.   Otherwise, mere technocratism will prevail, and man will be rectified of his core in the middle of material prosperity.   One way of improving this atmosphere is putting greater emphasis on moral education also on the university level.

In most universities, ethical education is confined to some basic or core requirement (if it is included at all) as something that has to be pulled off by the student in the freshman or sophomore year of studying, and then forgotten to move to a higher plane of major requirements and specialized courses.   Students have, therefore, a rather casual contact (if any) with ethics during their university life. Requirements for majors virtually never include a course projecting the field the students are majoring in onto a broader plane of social life. They virtually never force students to give a thought on the moral perspective of scientific activity proper to their domain.   Scientific and technological issues are, as a rule, divorced from social problems they can bring about.   In most cases, only philosophy and theology students are exposed to these aspects of science and technology; science and engineering students are not.

This situation is a token of short-sightedness on the part of the academia.   The universities, by narrowing the extent of moral educa-tion, stress the preference rampant in business and in military.   Most often than not, in these two latter domains moral issues are the last to count and the least in importance. But if any institution is to go against this trend, it ought to be the university. Therefore, the value of moral education should be revitalized and its importance elevated to the highest possible level. It may well be that it will not change everyone at once, but by the sheer presence of ethics related courses in all majors the university would emphasize an importance of moral issues and indicate that the scholarly and engineering activity must not be divorced from an attitude of social responsibility.

Speaking in practical terms, an ethics and social responsibility class should be included in every curriculum both at undergraduate and, especially, graduate levels, and in particular in medicine, science, and engineering curricula. The class should be oriented to a given domain, for instance, ethics and mechanical engineering, ethics and computer science, ethics and atomic physics. These courses should include some general discussions of ethics, but they should be geared to these particular areas by, for instance, considering some real or imaginary scenarios and discussing behavior of their participants. Examples of this kind of scenarios may be found in *Computer ethics* by Tom Forester and Perry Morrison (1990), in *Case studies in medical ethics* by Robert Veatch (1977), and in *The ethics of management* by LaRue T. Hosmer (1987). They could also include some real events of history, for instance, launching the atomic bomb project, or medical research on fetuses. In this way, the student would at least realize that including moral considerations in the case of scientific and engineering projects is as important (in fact, more important) than pondering the purely scientific and engineering aspects of these projects. In would also be a very constructive way toward, to use the words of physicist Edward U. Condon, "the joining of the goals of ethics and the goals of science" (Condon 1961, 22; cf. also Unger 1982, 138).

The university is much too concerned with just having students learn some specialized topics and with pouring into students' heads as vast an amount of technical and specialistic information as possible. Only undergraduate students receive some broader education, graduate students are usually confined to their field without any concern that this field is not an art for art's sake: it should serve the society. Therefore, the university should sensitize them to broader issues, to ethical questions, and to the problem of responsibility. Today, an effort on the part of the university in this direction is next to naught. It stresses the need for "pure and simple scientific curiosity". Such a curiosity is, indeed, a beautiful virtue, but it is rather egotistic and therefore, not full without including the needs of the others in the picture. It is the responsibility of the university to have students think about what this curiosity can lead to. But without even trying to show it, an implicit assent is given to possibly morbid consequences of this curiosity.

Most of the issues mentioned up until now are addressed one way or another in professional ethic codes. Practically every profession has an ethics code that spells out expectations with regard to behavior of professionals in a given field. They list some dos and don'ts specific to the field which serve as guidelines, especially in conflicting situations. The problem with these codes, however, is that they are not being used. The lofty goal to guide professionals is thwarted by the professionals themselves, since the codes are rarely referred to, and only infrequently used in daily routine work. One study showed that among a group of professionals of several disciplines, only less than 10% were able to find a copy of the ethical code in their office, which "supports the suspicion that possession of the code is a low priority for most practitioners" (Reeck 1982, 69). The situation may be defended by the statement that the professionals internalized all ethical rules so that no reference to any code is necessary. However, such an interpretation seems to be much too generous.

This low status of professional codes reinforces the need for moral education at the university level. If professional ethics is not a part of education and if establishing such a code is left to some ethics committees, then it should not be surprising that the status of these codes is not exceedingly high. But if teaching ethics, professional ethics in particular, is given the same treatment by the university as teaching other courses, then the attitude toward these codes may be expected to be improved.

It is clear that introducing ethics-related courses will not improve situation by itself and it will not automatically change the attitude of students. It may even worsen it, if the university does not do what it teaches, because the gap between theory and practice is an invitation to the view that ethical issues are completely irrelevant, and teaching them is a stilted way of wasting time. The university should at all times keep in mind that it is an educational institution promoting moral, social, and - to be sure - scientific growth of the students, and all its activities should be geared toward this goal. Otherwise, it would promote hypocrisy and justify practices which in theory it repudiates. One example are hiring practices.

It is at least disconcerting that not infrequently when university wants to hire a star scientist, the first bargaining is about having teaching load reduced. The university, an educational institution hires professors so that they do not teach. Other goals are more important, which in lofty terms are put as developing science, doing research, breaking new technological and scientific barriers. Therefore, gaining knowledge takes priority over educational goals. Scientific research has more clout than education, since publication of papers and books, and conference presentations are more acclaimed than good teaching. However, this attitude promotes as much egoism, self-centeredness and self-promotion as it furthers the development of science. In the long run, and certainly in the short run, the tendency is rather to the former than to the latter. What counts is *my* book, *my* discovery, *my* patent - which leads to promotion and raise - rather than the good of students, interest of university, and the community in general. This, to be sure, does not pass unnoticed. Students see that concentration on one's own research rather than on teaching and service to university is more acclaimed and, unwittingly or not, they adopt this attitude and perpetuate it. If science and technology is the greatest achievement of our century and if developing science is intertwined with concentration on one's own research to the exclusion of other obligations, then is there any escape from conclusion that remaining in one's own little circle of research is most important?

This unfortunate situation ought to change. University should not forget its mission: it is an educational institution first and foremost, and this mission has to be on everyone's mind when looking for new faculty. By releasing new faculty from teaching obligations, university promotes an attitude of avoiding to fulfill by an institution its basic mission as acceptable, even worth acclamation. The interest of scientific research replaces university's mission, and the situation is being sanctioned in which the students are for university, not university for students. What practical lesson can students learn from it?

Another important problem which faces the university is the way of funding university, in particular, scientific research, and the way the scientists draw pecuniary advantages from results of their scientific and technological research.

The search for truth is an expression of disinterest in theoretical activity, and of finding satisfaction in knowing more and in greater depth. The pursuit of the good is an expression of selflessness and of an interest in the well-being of the other. In both cases ego breaks a circle of its private world and opens itself to the depth of the universe and to the gamut of needs of the other. Such an ego enriches itself by this openness and finds the main (if not the sole) gratification in seeing that what has been achieved deepens our understanding of the world and also improves our life. However, if the underlying motive is a pecuniary reward, then science, sooner or later, turns into a market-place for which the truth is merely one of several possible means that, if expedient, can be replaced by something more convenient. As a result, we have quarrels between faculty and administration concerning 'how much' and 'to whom,' as in the case of software copyright.

There is always a danger of such quarrels if the university meets the market, if the results produced by the university lead to practical applications. Then there is a problem with to whom royalties belong, and who is the owner of the results. If a faculty member publishes a book, the author owns all rights to it and the university does not try to withhold them on account of the author's being employed by the institution. The reason for this is that, as a rule, the faculty do not use university's facilities or time to write. On the other hand, the university participates in the royalties received as a result of a patent, since the university's laboratories and time usually need to be used to conduct research leading to the invention (Hollander 1984, 222).

The problem becomes somewhat sensitive in the case of computer software, which is copyrighted, not patented; however, it is not treated on an equal footing with other copyrighted material, since universities desire to possess the copyright or at least to share it. "The real issue is money," Ivars Peterson comments on these developments. Books written by the academics rarely bring substantial income (with the exception of some textbooks). On the other hand, writing software is a lucrative occupation and the university would like to have its share. Besides, writing software frequently requires expensive hardware - owned by the university. Therefore, there is a problem of "substantial use of university resources." These resources tend to be more and more costly and the university tries to regain some cost by participating in the copyright and patent royalties. This is of course understandable; therefore just blaming "greedy administration and an overly restrictive copyright and patent policy" for "poisoning the atmosphere on a university campus" (Peterson 1985, 188) is one side of the coin - the other side is just as tainted by greed. And this greed results in certain tendencies that tarnish the image of science.

The danger of turning science into a marketplace was fully realized by great scientists to the extent that some of them did not even want to patent their inventions or discoveries, feeling that it is somewhat unbecoming for a scientist who finds fulfillment in the process of creating and discovering.

For instance, Benjamin Franklin refused to patent the lightning rod and to profit from it. He also declined to patent the fireplace, turning the model to his friend, Robert Grace, to manufacture saying "that as we enjoy great advantages from the inventions of others, we should be glad of an opportunity to serve others by any invention of ours, and this we should do freely and generously" (van Doren 1938, 172, 117; Franklin 1960, 419). Similarly, Louis Pasteur, who estimated that the cure for silkworm disease he discovered after two years of research would bring millions of francs to the growers, was not interested in patenting it since "'in France scientists would consider that they lowered themselves by doing so.' He was convinced that a man of pure science would complicate his life, the order of his thoughts, and risk paralyzing his inventive faculties, if he were to make money by his discoveries" (Vallery-Radot 1926, 129). This statement is seconded by Marie and Pierre Curie, whose newly discovered radium could have found industrial and medical applications. They did not intend to patent their discovery, and after receiving a letter from Buffalo, they responded by sending a description of the process of radium purification. Marie Curie contended that patenting the discovery "would be contrary to the scientific spirit ... Physicists always publish their research completely. If our discovery has a commercial future, that is an accident by which we must not profit. And radium is going to be of use in treating disease. It seems to me impossible to take advantage of that" (Curie 1940, 204). Interestingly, neither Pasteur nor the Curies tried to rationalize receiving a monetary reward for their discoveries because of the need to equip their laboratory, although this need was strongly felt by them.

This commendable attitude is, fortunately, not a matter of a romantic past; for instance, Richard Stallman distributes his *GNU* software system to everyone interested in using it; Donald Knuth released his excellent *Tex* as free public domain software, and public bulletin boards are being more and more expended with programs written by people interested just in writing and sharing them. It is not, therefore, true that copyrighting and patenting gives a necessary incentive to make useful and interesting inventions and discoveries. Actually, it may force potential discoverers to work on projects only because they may bring monetary rewards. This reason, to be sure, does not suffice to valuate an invention, since - as it is estimated - 75 % of all patents litigated to a contested and published decision are destroyed, and 75 % of granted patents are bogus inventions: useless, queer, fruit of wasted effort (Gilfillan 1964, 16-17, 95).

Admittedly, the attitude of Franklin, Pasteur, or Curie does not have a place in business, whose essence is not truth seeking, but profit production and immediate palpable results. But it should not be much of an exaggeration to see "the scientific spirit" at least in universities. There is, however, a danger of stifling or at least debilitating this spirit when it comes to research funding; there is a clash of interests between business and university, and the latter is often a source of funding. In these circumstances moral issues become especially acute, since rather infrequently ethical considerations may be suspended. In a Jack Magarrell's survey of nearly 300 chief executives in the private sector, one-third responded that the decision concerning university funding would hinge upon the faculty's economic and political views (Baron 1983, 158-9). What should one's ethical stance be in the face of such a statement? Hide the offending views to gain monetary support for the university? Suspend the scientific spirit? This problem is not restricted to the private sector; it may be found also in the case of the best (the most affluent, that is) sponsor of scientific research, which is the military.

The impact of the military and its oriented research upon academia is very well proven; for instance, Carl Barus (1987) shows that the curriculum of electrical engineering was shaped by the military. Besides, military funding of computer science research in the academia is increasing; in 1983, 60 % of federal funds came from the Department of Defence and in 1984, 40 % of all computer related research in the nation was funded by DoD (Winograd 1984, 706). As a sponsor, the military directs the researchers into specific areas. One of the fields well funded by the military is AI. It is even asserted that "without the constant military funding that it has received over the last 30 years or so, AI would almost certainly be just a quaint academic curiosity which few people would have heard of and [in which] even fewer were interested and studying" (Forester, Morrison 1990, 131). Is it then a genuine interest that leads researchers to the field of AI, or are they drawn to this field after learning that it will receive a lavish funding from the military? Are the researchers enticed into this particular field because they can see an opportunity for testing their theoretically obtained results, or an opportunity for receiving a grant? In the light of the quoted opinion the question seems to be rather rhetorical. The fact of determining research by "where the money is" truly "degrades both science and scientists" (McCain, Segal 1977, 165).

If it is not enough for the origins to be especially savory, the motives behind some scientific discussion may turn out also to be rather unsavory. For instance, Seymour Papert remarks, tongue in cheek, there was a tension between "natural sister" and "artificial sister", that is, between "connectionists and programmers" concerning "the access to Lord DARPA's research funds. The natural sister had to be slain." As a result, the book *Perceptrons* was published that showed limitations of simple neural nets. "By 1969, the date of the publication of *Perceptrons*, AI was not operating in an ivory-tower vacuum. Money was at stake" (Papert 1988, 3, 7). As a result, the book was embraced by the AI community and served very well as a tool of suppressing the development of connectionism for many years, since most of writing concerning perceptrons "is without scientific value" (Minsky, Papert, 1969, 21). The recent developments in computer science prove that there is some scientific value in connectionism, after all.

And this again leads us to the fundamental question of the sense and meaning of science and scientific endeavor: Is science to be interested in the pursuit of truth, one of the three classical virtues, or is it to be at the service of a funding agency? Do scientists want to go deeper and deeper into their subject matter because they deem it to be important, revealing, and promising in its own right, or do they want to give their time and endorsement to research which is well funded but maybe not so important? Is the primary goal of science to widen our under-standing of the world or be a servant of rich sponsors? These questions are all the more important because the researchers referred to are also university professors, hence educators, hence authorities for students, hence examples of the correct attitude toward science. Is it ethical then, especially for leading universities, to use such funding as a leverage to entice students not only to these schools, but to a particular type of research that otherwise might not stir any interest at all? Is it ethical to channel students - directly or indirectly, mainly by example - to certain areas on account of the fact that they are well funded? What should be the position of university researchers in this debate?

It seems that most faculty members do not want to be associated with work servicing the military. However, when it comes to obtaining or not obtaining a contract, moral scruples are oftentimes set aside. This leads to work on projects that otherwise would not incite much, if any, interest among researchers. What counts is a grant and the subject matter is less important, even less the sponsor. There are researchers who reject any type of military funding. Generally, women scientists may be expected to undermine the priority of the military-oriented research in their recommendations for national policy concerning funding scientific research, since "women consistently register their preference for having more money spent for health, education, and social welfare programs and less money spent on defense and the military" (Rosser 1990, 41). Some universities accept such funding only for basic research. Other universities avoid any type of military funding that may lead to classified research, blocking presentation of papers, even to restrictions concerning admission of foreign students. David Parnas feels that the professionals can participate in military funded projects that are effective and do not go "beyond the legitimate needs of the country." But he hastens to add that "too many do not ask such questions. They ask only how they can get another contract" (quoted in Dunlop, Kling 1991a, 659). This invokes in memory an answer given by the Emperor Vespasian to his son Titus after instituting a tax on public urinals; Titus felt that it was unseemly and castigated his father, who, in response, held money under his son's nose and said, *pecunia non olet*, money does not smell.

Those who disagree with the Vespasianian attitude do not embark on a project only because it is funded but that may serve purposes far removed from what scientists find non-objectionable. They do consider on moral grounds the merit of a project being funded, its rationale and usefulness, and try to promote a project they regard valid and important, facing the prospect of being cut off from a lucrative source of funding. In an artless statement, the head of research for the Pentagon said that if university professors "want to get out and use their roles as professors to make statements, that's fine, it's a free country ... But freedom works both ways. They're free to keep their mouths shut ... and I'm also free not to give them money ... I have a tough time with disloyalty" (quoted in Winograd 1984, 713).

Fortunately, it is not the only feeding tube available, so that the scientists are not forced to put aside moral qualms at the cost of conducting research. First, they are obliged to seek truth, and teach their students how it can be discovered. Second, they have an obligation to cultivate moral values in the realm of science. If these values are suppressed in order to advance progress in science then priorities are ill-defined. If this progress leads to a breakdown in the moral fiber of scientists, and, consequently, of the public, then it is better not to have it at all. Progress is often, all too often, identified with technological progress, with the capability of doing things more swiftly, more cheaply, more efficiently, and in larger quantity. But if these goals are not undergirded by moral values, by having the well-being of humanity as the primary goal in mind, then technological progress turns into a swift, cheap, and efficient killing machine. Therefore, there is very little truth in the claim that in order to conduct some kind of research, scientists - turning a blind eye on moral issues - have to turn to an otherwise unwanted sponsor, because this can make the research possible.

Arguments that national defense requires work for the military, albeit appealing, are not convincing. A recourse to such arguments simply means that other options of enhancing national defense are less viable, or not valid at all, and that the primary principle of enhancing it is summarized in Vegetius' maxim, *qui desiderat pacem, praeparet bellum*, let him who wants peace prepare for war. Yet this is not necessarily true, and many people believe that the sword is not the only way of solving problems, especially in the global arena. An armed conflict is not inevitable, and the military buildup is not the only means of ascertaining it. In the long run, diplomatic and economic means, although less spectacular and more mundane, are much more efficient.

Thus, when it comes to a conflict of priorities, the academia should hold fast to its traditional values, even if in the short run it may lead to some financial difficulties. The academy has an enormous responsibility for the education of students and for instilling in them what is really important and lasting: respect of truth and well-being of humanity. These values cannot be attained at any cost, since the way of attaining them is a part of the result. Therefore, if funds are needed to better equip laboratories, and if obtaining these funds is connected with temporary suspension of the university mission, then the university should think twice before making decision. *Pecunia olet*, money does smell, and it may result in irreversible damage if the academia turns into a money-making institution. *Pecunia olet*, and the smell may poison the atmosphere to the extent that the very purpose of the existence of the university will become sullied. The university has to defend itself against commercialization, and discriminate when it comes to research funding; the latter should be the means of achieving the main goal of academia: showing students in theory and practice what is science and how it should be treated. However, the end does not justify the means. The academy should be the first place where this adage is put into practice, and the last place to follow a recommendation of an AI researcher who says that "Yale should harness its professors to make money" (Kendig 1983, 36). It is truly regrettable that such statements are made by academicians. What Yale needs is to rekindle the scientific spirit of Franklin, Pasteur and the Curies. And not just Yale, but academia as a whole. The academy should also be a place where all the faculty, not only exceptional individuals, ought not to be "middle managers" and "instructional leaders" but "educational leaders who embody the ideals and exemplify the virtues which we aspire to as an educational community" (Grant 1988, 227).

References

Baron Robert 1983, Higher education and the corporate sector: Ethical dilemmas, in M.C. Baca, R.H. Stein (eds.), *Ethical principles, practices, and problems in higher education*, Springfield: Charles C. Thomas Publisher, 153-164.

Barus Carl 1987, Military influence of the electrical engineering curriculum since world war II, in Dunlop, Kling 1991, 717-727.

Blasi Augusto 1985, The moral personality: Reflections for social science and education, in Berkowitz Marvin W., Oser Fritz (eds.), *Moral education: Theory and application*, Hillsdale: Erlbaum, 433-444.

Bloy Myron B. 1966, Technological culture, the university, and the function of religious faith, in Brickman W.W., Lehrer S. (eds.), *Automation, education, and human values*, New York: School and Society Books, 328-341.

Boorstin Daniel J. 1991, History's hidden turning points, *U.S. News and World Report*, April 22, 52-65.

Bronowski Jacob 1955, *Science and human values*, New York: Harper 1965.

Condon Edward U. 1961, The challenge of science to human values, in Johnston E.G. (ed.), *Preserving human values in the age of technology*, Detroit: Wayne State University Press, 7-26.

Curie Eve 1940, *Madame Curie*, New York: Doubleday.

Dewey John 1922, *Democracy and education*, New York: Macmillan.

Dewey John, McLellan J. 1895, The psychology of number, in John Dewey, *On education: Selected writings*, New York: Random House 1964.

Dunlop Charles, Kling Rob (eds.) 1991, *Computerization and controversy: Value conflicts and social choices*, Boston: Academic Press.

Dunlop Charles, Kling Rob 1991a, Ethical perspectives and responsibilities, in Dunlop, Kling 1991, 654-63.

Ellul Jacques 1990, *The technological bluff*, Grand Rapids: Eerdmans.

Evans Christopher 1979, *The micro millennium*, New York: Washington Square.

Forester Tom, Morrison Perry 1990, *Computer ethics: Cautionary tales and ethical dilemmas in computing*, Cambridge: The MIT Press.

Franklin Benjamin 1960, *The papers of Benjamin Franklin*, New Haven: Yale University Press, vol. 2.

Gilfillan S. Colum 1964, *Invention and the patent system*, Washington: U.S. Government Printing Office.

Grant Gerald 1988, *The world we created at Hamilton High*, Cambridge: Harvard University Press.

Hollander Patricia A. 1984, An introduction to legal and ethical issues relating to computers in higher education, *The Journal of College and University Law* 11, 215-232.

Hanson Dirk 1982, *The new alchemists*, New York: Avon.

Hersh Richard H., Paolitto Diana P., Reimer Joseph 1979, *Promoting moral growth: From Piaget to Kohlberg*, New York: Longman.

Hosmer LaRue T. 1987, *The ethics of management*, Homewood: Irwin.

Jarrett James L. 1991, *The teaching of values: Caring and appreciation*, London: Routledge.

Jastrow Robert 1981, *The enchanted loom*, New York: Simon & Schuster.

Kendig Frank 1983, A conversation with Roger Schank, *Psychology Today* 17, April, 28-36.

McCain Garvin, Segal Erwin M. 1977, *The game of science*, Monterey: Brooks/Cole.

Minsky Marvin, Papert Seymour 1969, *Perceptrons*, Cambridge: MIT Press.

Maslow Abraham H. 1971, *The farther reaches of human nature*, New York: Viking.

Papert Seymour 1988, One AI or many?, in Graubard S. (ed.), *The artificial intelligence debate*, Cambridge: MIT Press, 1-14.

Parnas David L. 1985, Software aspects of Strategic Defense Initiative, in Dunlop, Kling 1991, 593-611.

Peterson Ivars 1985, Bits of ownership, *Science News* 128, 188-90.

Reeck Darrell 1982, *Ethics for the professions*, Minneapolis: Augsburg.

Rollin Bernard E. 1991, Some ethical concerns in animal research: Where do we go next?, in Baird Robert M., Rosenbaum Stuart E. (eds.), *Animal experimentation: The moral issues*, Buffalo: Prometheus Books, 151-8.

Rosser Sue V. 1990, *Female-friendly science*, New York: Pergamon Press.

Rotblat Joseph 1985, Leaving the bomb project, *Bulletin of the Atomic Scientists* 41, 16-19.

Siemens Werner 1886, The ethical influence of machinery on labor, in Burke J.G. (ed.), *The new technology and human values*, Belmont: Wadsworth 1966, 18-24.

Simon Sidney B., Robert D. O'Rourke 1977, *Developing values with exceptional children*, Englewood Cliffs: Prentice-Hall.

Unger Stephen H. 1982, *Controlling technology: Ethics and the responsible engineer*, New York: Holt, Rinehart & Winston.

Vallery-Radot René 1926, *The life of Pasteur*, Garden City: Doubleday.

van Doren Carl 1938, *Benjamin Franklin*, New York: The Viking Press.

Veatch Robert 1977, *Case studies in medical ethics*, Cambridge: Harvard University Press.

Winograd Terry A. 1984, Strategic computing research and the universities, in Dunlop, Kling 1991, 704-16.